Facing the Promotional Interview

By John Mittendorf

Copyright© 2003 by
PennWell Corporation
1421 South Sheridan Road
Tulsa, Oklahoma 74112-6600 USA

800.752.9764
+1.918.831.9421
sales@pennwell.com
www.pennwellbooks.com
www.pennwell.com

Supervising Editor: Jared d'orr Wicklund
Production Editor: Sue Rhodes Dodd
Book Designer: Beth Caissie

Library of Congress Cataloging-in-Publication Data

Mittendorf, John, 1940-

Facing the promotional interview / by John Mittendorf.-- 1st American ed.

p. cm.

Includes bibliographical references and index.

ISBN 1-59370-002-4

1. Fire extinction—Examinations—Study guides. I. Title.

TH9157 .M578 2003

628.9'25'076--dc21

                                                           2003011004

ISBN 1-59370-002-4

Printed in the United States of America

1 2 3 4 5 07 06 05 04 03

Facing the Promotional Interview

This book is dedicated to the numerous chief officers who unwittingly had to sit on my various promotional interview boards and listen to my feeble attempts at expressing myself. If they were to read this book, they would probably wonder why I didn't *practice* what I *preach*.

Also, to my sister, Linda, who unknowingly was responsible for putting a pencil in my hand and starting my literary career, and ultimately, making this book possible.

Contents

Foreword

Being asked to author a book on promotional interviews for the fire service is a humbling and rewarding experience. First, from the perspective of humbling, what makes a person an "expert" in the promotional interview process? I would use the word *insight* as a replacement for the word *expert*. The interview picture becomes somewhat clearer when a person has the opportunity to *sit on both sides of the interview table*. In my case, I never really understood the interview process and its subtle requirements until I had the opportunity to evaluate and grade interview candidates (both entrance and promotional). It was this combination of experiences and watching numerous candidates nervously *go through the motions* in connection with my failures and successes that are incorporated into this book.

Second, the rewarding part of writing this book is incorporating the aforementioned experiences (some good, some not so good) into a single resource that hopefully will allow the reader to maximize promotional interview results and avoid common pitfalls. There is no greater joy to any fire service author than to meet someone who has read your book and receive the comment, "thanks, the stuff in your book worked for me." In the case of promotional interviews, hopefully you will only need to use the information in this book one time per promotion!

In conclusion, an important note of thanks to Francie Halcomb and Jared Wicklund of *Fire Engineering*, who persisted in the development of this book, and to Paul Bunch and Steve Engh for the illustrations.

1. Priorities before the Interview

Introduction

Of all the challenges employees of the fire service encounter during their career, the ability to be *first-in* to a structure fire, and put the *wet stuff on the red stuff* is potentially the most dangerous, but definitely the most exciting. Particularly if the fire was in a district other than theirs. However, the most enduring experience employees of the fire service will ever encounter is the opportunity to promote to a higher rank and/or additional responsibility for the following reasons:

- You will either succeed or fail in your endeavor. Remember, some memories can last a lifetime. Make them positive and memorable.

- Your success or failure will impact your standard of living for the rest of your career and lifetime. That also includes retirement.

- Promotional opportunities (i.e., interviews) are normally accompanied by periods of high stress levels. Blood pressures of 320/200 can be easily attained and maintained for brief periods.

Therefore, promotional opportunities from a fire service perspective can be defined as *a life- and career-changing opportunity*. Let's now add one more ingredient to the aforementioned definition—you. Without a doubt, you are the only person who can significantly control the type of reputation you develop during your career. Additionally, you are also the only person who has total control over the amount of time and dedication devoted to achieving your interview goal and the type of presentation you deliver during your interview (crunch time). You and only you determine the success or failure of the preceding factors. The next seven chapters will be based on the following five key thoughts:

1. Are you serious, or do you enjoy taking interviews?

2. Interviews can be fun, because your future depends on (only) you.

3. Applied elbow grease is the key ingredient in your success.

4. Nothing can be substituted for proper preparation.

5. First impressions are lasting impressions.

Effective Preparation

If you met someone for the first time, and that person impressed you as being able to express himself or herself distinctly, clearly, and unmistakably, you would probably classify that person as articulate. Similarly, if you took an interview examination for a promotional position and received a score of 98, you would definitely be classified as having the key to effective interview presentation, or else as extremely lucky. For this book, however, let us discount the somewhat undependable element of luck, and focus our attention on some vital keys that will assist you in preparing and delivering an effective interview presentation for a typical civil service interview promotional examination.

As a point of interest, the definition of effective is *producing a desired effect*. The effect you *want* to produce will be governed by how serious you are about a promotion and the time you are willing to invest in preparing for that promotion. The effect you produce will be governed by a combination of your natural ability, desire, and degree of preparation. Obviously, natural ability is not easily modified or changed. However, desire and the degree of preparation is easily modified or changed. This is accomplished by the time and *proper effort* you are willing to devote to preparing for an interview. Notice that the phrase "proper effort" was italicized in the last sentence. The focal point is the word *proper*. A musician will readily admit that practice does not make perfect. Perfect practice makes perfect. In the fire service, we are constantly challenged to *train the way we fight*. So, when preparing for your promotional interview, prepare correctly as if your future depends on it, because it does!

Interestingly, desire, degree of proper preparation, and confidence are directly related. As the degree of preparation increases, your desire to promote yourself will also increase. Additionally and sometimes most importantly, the confidence that you can effectively accomplish that task will also increase. Let me restate the last sentence from a different perspective. Effective interview preparation and presentation takes work, and lots of it. The ability to sell a product is dependent on a total knowledge and belief in and of the product (you), and a thorough knowledge of the customer (the interview board). If you are not sold on your capabilities and level of performance, you also will not convince interview board members you are the candidate they are looking for.

If you are serious about preparing for an interview, then passing the interview is not your primary goal. Your sole focus is convincing the interview board that you are the top candidate. Period! Your grade will reflect your effectiveness in accomplishing that goal. If you are number two on the promotional list and the person ahead of you is the only candidate chosen during the life of the promotional list, then you obviously were not successful in obtaining a promotion. However, you may have been very successful in adding to your cache of life's experiences. Remember the famous saying, "There is always enough time to do it right the second time."

Effectively preparing for an interview is not accomplished in a short period of time. Some candidates spend months preparing for a written examination but only a few days, weeks, or months for an interview that may be worth 50% of the final score. As a prime example, after conducting a workshop on promotional interview techniques, I often receive a phone call inquiring if I have any interview study materials. If I answer in the affirmative, the next question is often, "Can you overnight the material to my home?" or "How soon can I get the material?" As the overnight charge is higher than the cost of the material, I always ask, "What is the importance of overnight delivery?" The answer is virtually always the same, "My interview is in several days!" In connection with the lack of apparent preparation, what do you think the potential for success is with this approach? Do you think this type of candidate will be able to change his/her presentation style in several days?

Balanced Approach

Unless a candidate is a naturally *gifted* speaker (which is extremely rare), the amount of time that a candidate intends to devote toward a promotional examination process should be balanced between studying for the written examination *and* preparing for the interview (Fig. 1–1). This will allow a candidate to master the technical knowledge necessary for a written examination and develop the experience, confidence, and presentation techniques that are necessary for an interview. Remember that regardless of how well a candidate performs on a written examination, it is generally the interview (or assessment center) that determines your future job position. From an analogy perspective, assume you are about to take a ride in a train. Your performance on the written examination will purchase your ticket, and your interview performance will determine where you will sit on the train (at the head of first class, in the middle of the coach, or at the end in economy).

Fig. 1–1 Develop a Balanced Study Program

Let's overview a balanced approach toward preparing for a written and interview examination process (Fig. 1–2). First, assume the written examination is scheduled for August 2004, and the interview is scheduled for November 2004. Next, also assume you want to devote 10 months toward preparing for the process (written examination and interview). Also remember that life does go on while studying for a promotional examination. This means you will start your study program during January 2004. However, before starting your study program in January, consider the following issues.

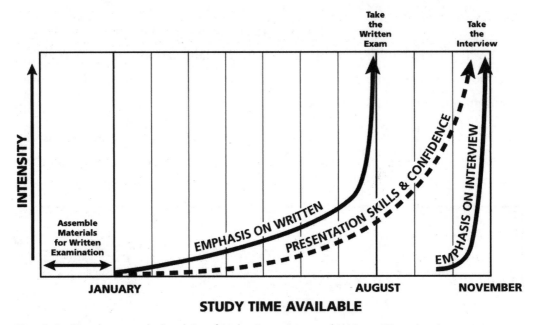

Fig. 1–2 Simultaneously Studying for the Interview and Written Examination

Job description and announcement bulletins

If possible, obtain a copy of the job description and announcement bulletins for your anticipated promotional examination. Every fire department and/or personnel department has a job description for every job position in your fire department. Job descriptions overview the elements (a road map) of each job position. This is extremely important and a key starting point for your study program. Job announcements overview the areas of responsibility that will be examined. As an example, the National Fire Protection Association (NFPA) Handbook for the Los Angeles City job announcement for captain lists the specific chapters that can be on the examination. Using a job announcement can keep you from wasting valuable study time on material not relating to the specific job description. This is a consideration that you must determine before you assemble your study material.

Additionally, job announcement bulletins usually indicate the *weighted* (i.e., written examination 50%, interview examination 50%, etc.) portions of the examination process. For example, assume the written portion of your promotional examination process will comprise 40% of the total grade, and the interview will comprise 60% of the total grade. How important is the interview? Additionally, as long as we are going to consider the requirements for a balanced study program, remember that a 70 on the written portion in connection with a 92 on the interview does not normally produce a winning final score! The job description and announcement bulletins set the foundation for determining the appropriate material that should be encompassed in a balanced study program.

Assemble material

Prior to the January date (Fig. 1–2), take the time that is necessary to assemble your written study material. You should have a copy of *all* the written material that you will be responsible for. This may include your fire and building codes, any appropriate specialized books (International Fire Service Training Association [IFSTA], *NFPA Handbook, Fire Officer's Handbook of Tactics, Truck Company Operations*, etc.), your department's manual of operations and/or rules and regulations, and so on. This can mandate the purchase of any necessary books and the copying of your department library. This will allow you to highlight your own material, make notes, and be able to maintain the proper context of the original material while studying. Although study summaries, reference material, and self-help materials can be helpful, you should primarily study the basic material in your department library. Remember, the person preparing the written examination is reading the identical material from your department library.

Study partners

Decide whether you will study alone or with several partners. Study partners who are equally serious about the examination can enhance the study process for the following reasons:

- Multiple partners provide incentives and accountability during the study program. It takes a great deal of self-discipline to study a portion each day for extended periods when your pals are having fun relaxing.

- Multiple partners can effectively cover more material than a single person. This is accomplished by assigning various sources of information to each person that can be researched and summarized for other members in the study group. I prefer to assign peripheral material (specialized books, fire codes, *NFPA Handbook*, etc.) to individuals.

For example, assume your examination mandates you are responsible for studying the *NFPA Handbook* and your fire code. Assign one person to cover the first half of the handbook, another person to cover the last half, and a third person to cover the fire code. This will allow these members to give a timely overview of these materials while all study club members simultaneously study the *meaty* stuff (your department rules and regulations, department directives, etc.).

When all the study material has been assembled, begin your study program (January in Fig. 1–2). Initially, set aside appropriate time each day for study. Remember to start slowly and let your capacity for concentrating on the material naturally accelerate during the study program. It is virtually impossible to sit down and start studying for several hours without your mind wandering (remember your pals are having fun relaxing). When your mind wanders, George Washington and you have one thing in common; you are both history. Therefore, allow your mind to adapt to a radical change in focus and concentration.

Initially, all material should be read through at least once and preferably twice to

- Develop an overview of the material to be studied.

- Indicate what portions of the material need specific or additional emphasis. Study material can be separated into the following categories:

   ✓ Material that needs to be learned

   ✓ Material that is "somewhat" familiar

   ✓ Material that is currently known

Additionally, this author suggests making typewritten notes on difficult, important, or unknown material and placing them in a notebook for easy and frequent reference. This is called *force feeding*.

When the material has been separated and/or categorized as previously discussed, appropriate time can be delegated to each category. This will provide the best use of study time. The following are general written examination study tips:

- Spend the maximum amount of time on material you don't know and the minimum amount of time on material you do know.

- Spend the maximum amount of time on material that can be expected to yield the most questions and the minimum amount of time on material that yields minimal questions.

The preceding two points can often be determined from several sources as detailed in the next section.

- If available, look at the announcement bulletin and past exams to determine the areas questions are drawn from (it is amazing how predictable personnel departments can be when writing a promotional examination as they usually look for the easy route). Remember, most personnel departments are responsible for producing promotional examinations for all city departments within your municipality, not just your fire department. These questions can also provide an excellent source of review material.

- Difficult material can be placed on 3 x 5 cards and reviewed while riding to work, etc. Remember that 3 x 5 cards can be a time-consuming process, and the information on the cards must be in the same context as the reference material.

- Stay current with changes to your department's library. New changes are excellent sources for exam personnel to access for examination questions.

- If essay questions will be used, the personnel who grade these "literary masterpieces" are normally looking for specific items that are predetermined by the personnel responsible for the exam, not the length of your answers. Therefore, practice writing essays on potential subjects, preferably in outline form. This style is easier and faster for you to write and for them to grade! To the average person, writing an essay in longhand requires more time and is usually more difficult to complete than outline form.

• As you initially begin to answer the questions on the written examination portion of the examination process, it is vitally important to carefully monitor your time so you are not suddenly faced with the prospect of having to answer 20 final questions in five minutes (Fig. 1–3).

Fig. 1–3 Monitor Your Time during the Written Examination Time Constraints

- Remember the following considerations:

  ✓ First, quickly look over the entire examination to get a feel for the questions and their layout.

  ✓ Next, go through the examination and answer the questions that are easy to answer. Do not initially spend precious time on a difficult question.

  ✓ Next, again go through the examination and answer the remaining questions that are easy to answer; continue to repeat this process.

  ✓ If there are several difficult questions, they will be the last questions you need to answer, and you can then utilize the remaining time to answer them.

  ✓ Do not spend valuable time trying to answer several difficult questions while you have not answered questions that are relatively easy.

  ✓ If you need to absolutely/positively guess at a question, exam writers favor the choice of *c* most of the time. It is recommended that if you can review previous examinations from the person who is writing your examination, determine what choice is favored (a, b, c, or d) and put that in your mental memory bank for potential future use. Remember, if you have to guess, this may help you to guess intelligently.

- If you are using multiple study partners, meet periodically to review material, discuss new material, and provide accountability to the partners. This author prefers every other week when starting and then weekly as necessary.

• With multiple study partners, it is highly recommended that the partners spend some quiet study time together several weeks before the written examination. This provides an opportunity to review all material in an environment that is free from phone calls (dump your cell phones), children, and other similar distractions. Prior to the study week, have an agenda for the week and follow it!

For example, this author attended a study week at a rented cabin in the mountains approximately two weeks before the written examination for captain. Thirteen members of the study club met for eight days. During that time, we reviewed all of our material, asked one another other questions on the material, and briefly discussed interview techniques. We only left the cabin for short walks and food. End result: all study club members were promoted to captain.

• Lastly and maybe most importantly, if you are married, ensure that your spouse understands and supports your commitment to prepare for a promotion. If successful, you both will profit.

Putting the Program Together

Once your study program for the written examination has commenced, I suggest you make a radical departure from conventional study programs. Typically, most candidates will study for the written portion only for a predetermined amount of time, and then take the written examination (January to August, Fig. 1–2). After the written examination, the typical candidate will take a break, and then delay studying for the interview portion as long as possible (August to November, Fig. 1–2). The reason for this is nobody enjoys placing themselves in a weak or embarrassing position/environment. If you are new to this game, take a practice interview to experience the full impact of the last sentence.

The delay tactic is commonly utilized, which compounds the initial problem. Remember, you do not develop razor-sharp speaking skills overnight, and the longer you prolong enhancing your speaking skills, the worse the problem becomes. Therefore, to put yourself ahead of the pack (way ahead of the pack), I recommend that you should also begin practice speaking (presenting yourself) in front of people on a slowly accelerating program while you study for the written portion of the examination (dotted line, Fig. 1–2). Notice that the emphasis in that last sentence was not on collecting additional study material and talking about interviews with other "experts." The emphasis was directly focused on you learning to effectively present yourself and developing the skills and confidence to accomplish that task. As previously mentioned, unless you are a naturally gifted speaker, you need practice and lots of it!

There are numerous methods that can be used to increase your interview presentation skills without initially making this concept a major project. Giving training programs in your department, becoming involved in social groups, enrolling in a community college speech class, or joining a local Toastmasters club will give you an opportunity to overcome nervousness, learn to think while standing on your feet before a group, and be able to organize and present your thoughts in a confident manner. Can you see the benefit of placing yourself in the same type of environment before your interview so you will be more comfortable during the interview?

Toastmasters is a national organization designed to produce positive results in strengthening speaking skills by offering an opportunity to regularly practice before people who are striving towards a common goal—effective public speaking. Additionally, conducting a drill in front of your fellow firefighters is difficult if you are not prepared.

A slowly accelerating speaking emphasis is designed to allow you to speak before people during your written examination study program. As previously mentioned, this will give you experience in thinking on your feet, and assist in learning to organize and logically present your thoughts from a foundation of confidence. Remember to begin slowly at first, even though it is difficult (one of the major reasons why most candidates postpone practicing for an interview until the last possible moment) and then increase your commitment as necessary. Initially, your presentation focus should be on the simple concept of *proper practice* and being comfortable in front of people. As the written examination approaches, begin to direct your interview presentation time toward fire department–related material and your pending interview. When the written examination is over, you can then direct your efforts toward fine-tuning and polishing your interview presentation skills (emphasis on interview, Fig. 1–2) instead of finishing the written examination and then trying to develop into a "great communicator" in a short period of time like the rest of the pack.

The following are suggested methods for practicing and improving your communication skills for your interview:

- Perfect practice makes perfect! If you continue to practice without correcting your mistakes or settle for less than your best, then you will deliver less than your best in your interview.

- Initially, put typical questions on 3 x 5 cards. Use these cards either in a specific order or randomly to practice spontaneity and develop your thoughts on various subjects. Using cards will develop a "base" foundation from which you will steadily improve.

- When you feel ready (or comfortable), practice in front of your spouse, study partners, members of your family, etc. Remember, it is more difficult to convince your friends how wonderful you are as opposed to people who don't really know you.

- Schedule practice interview boards with members of your department, preferably at least one rank above your present level. One word of caution—when you sit at the feet of "interview experts," you will be introduced to numerous styles and recommendations on how you should *do it*. Carefully consider each recommendation for what it is—an opinion. Develop your own style not someone else's.

- Without a doubt, one of the most powerful and effective tools to improve your speaking skills is to videotape your practice sessions. It is easy to take some 3 x 5 cards and have your spouse/family members ask you questions while the video camera is on. I guarantee that your initial attempt will make you wonder if your speaking skills are really that bad. Most likely, they are! However, the voice, mannerisms, and presentation style you will see and hear are yours. What you see is also what the interview board members will see.

This style of media allows you to specifically focus on weak portions of your presentation. However, I also guarantee that as you continue to tape your presentations, you will build on your strengths and reduce your weaknesses. As you improve, so will your confidence. Guaranteed! With practice, there is no reason why you should not be able to walk into the *real* interview with confidence. Although you may be nervous, you can still sit down in the hot seat, look each interview board member in the eye, and mentally say, *let's go!*

Summary Checklist

1. The definition of effective is *producing a desired effect.*

2. The effect a candidate produces in an interview is governed by the type and degree of time invested in preparation.

3. As the degree of preparation increases, so does desire and confidence.

4. If a candidate is not sold on his/her capabilities and level of performance, the candidate will not convince the interview board he/she is the candidate for the job.

5. The amount of time a candidate devotes towards a promotional examination process should be balanced between preparing for the written examination and an interview.

6. The goal of an interview candidate should not be to pass the interview but to convince the interview board the candidate is the best qualified for the job.

7. The interview generally determines a candidate's success in achieving a promotion.

2. Anatomy of an Interview

Introduction

After navigating through the basic elements and challenges of chapter 1 that primarily focused on the concept of improving your public speaking skills in connection with self-confidence (which are not overnight projects), let's shift gears. Consider the primary elements of an interview before we dovetail chapter 1 with the concepts of chapters 3 and 4. A promotional interview (or any type of interview, civil service or civilian) is dissected and summarized into two basic components—interview structure and interview concepts.

Interview Structure

Within the time constraints that an interview is limited to, the interview can be divided into three main portions (Fig. 2–1). Assume an interview is between the normal time frame of 30 to 40 minutes long. The first 5 minutes is normally known as the *opening* portion, the specific answer and question portion is the *body* or main portion of an interview, and the last few minutes when the interview is concluded is known as the *closing*.

Fig. 2–1 Interviews Can be Divided into Three Main Portions

Opening

Prior to you entering the interview room, the board will briefly read your application and resumé (if available) to get a basic idea of who you are and review your background. However, once you enter the room, it's show time. Initially, the board will try to get you as comfortable as possible by asking you a few questions, "What can you tell us about yourself? What is your background?" and other similar questions. This portion of the interview gives the board an opportunity to meet you, get the interview started on a positive note, and develop an initial impression of you.

This is one of the most important parts of an interview even though it is extremely subtle. Interestingly, it is also a part of your interview that you have complete control over. It is a fact that board members are human and will form an initial opinion of a candidate in a short period of time (Fig. 2–2). You are probably wondering, "What is the definition of a short period of time?" An impression or opinion of a candidate is formed within 30 seconds to 3 minutes from the start of an interview. Surprised? Ask any person who has had an opportunity to sit on *the other side of the table* how long an initial impression takes. The common answer is often *30 seconds to 1 minute flat.*

Fig. 2–2 Interview Boards Quickly Form an Opinion of a Candidate

In practical terminology, that means the first several minutes of an interview will give the interview board members a general idea of your final grade. Unless you fall asleep and hit the floor or accelerate like a Saturn booster rocket (Fig. 2–3), your initial impression will not significantly vary throughout your interview. Do you think the visual and verbal image you initially project is dependent on your preparation before your interview? This portion of the interview sets the foundation and direction for the balance of the interview. Luckily the wise and prepared candidate can determine the outcome of this portion as detailed in chapters 3 and 4.

Fig. 2–3 Be Prepared for the Initial Portion of an Interview

Body

Once the interview board has given you a chance to be somewhat relaxed and given you a few minutes to talk about yourself, they will shift gears and use the major portion of the balance of the interview to discuss various types of questions within specific parameters designed to accomplish the following goals:

- Demonstrate and/or expand on your qualifications.

- Analyze your thought process.

- Determine your knowledge of your department's rules, regulations, policies, procedures, etc.

- Evaluate your opinion on pertinent fire service subjects.

- Give you an opportunity to *take a stand* or *make a decision* on relevant issues/subjects.

The preceding types of questions are dictated by categories (normally seven to nine distinct categories) commonly predetermined by a personnel department. Some examples of these are interpersonal relationships, communication skills, fire prevention, decision-making ability, etc. The goal of the board is to ask you a question from each category within the time constraints of the interview process. Remember, a typical interview may only last for 30 to 40 minutes, yet this is the time allotted to get to know you, determine your qualifications for the new job, estimate your potential success in the new job, and rate your degree of success with the other candidates. The body portion of an interview provides most of the information to achieve these requirements.

Closing

Just prior to the end of an interview, the board will conclude the interview by indicating they are finished with their questions. Generally, most interviews are concluded with a phrase similar to "Do you have anything else you want to say or add?" When the interview board reaches this part of the interview, they are finished and are anxious for you to "hit the bricks" so they can evaluate your performance, determine your grade, factor your rating with previous candidates, and prepare for the next candidate. Since the final question offers the candidate several options, chapter 6 will expand on the available responses.

Interview Concepts

With the preceding overview of the three main portions of an interview, let's turn our attention to what an interview board is primarily looking for in a candidate and discuss four types of concepts that will be utilized and evaluated during an interview.

1. Interview boards must ask three basic questions.

Generally civil service rules require interview boards to ask three basic questions during the interview period. These questions can take many different forms but still must be asked to enable the board to evaluate the following:

- *What are your qualifications and how are you qualified for the new position?*

  It is difficult to be rated unless your qualifications, abilities, background, and other similar traits are communicated to the interview board.

For example, assume General Electric is interviewing for a person to build electric motors. During the interview, do you think the board will question you on your qualifications and abilities necessary to build electric motors? Absolutely! Civil service interviews are no different. However, it is interesting that this concept is frequently hidden in subtle questions such as:

✓ "What can you tell us about yourself?"

✓ "Can you give us an overview of your background?"

✓ "How have you prepared for this promotion?"

✓ "Where have you worked?"

- *What is the new position for which you are interviewing?*

It is difficult for an interview board to rate your knowledge of the position you are interviewing for unless you are given the opportunity to display your knowledge of the new position. Amazingly few candidates can succinctly define the position for which they are applying. Before you ever step into a promotional interview, ask yourself the following question, "What is the definition of the position for which I am applying?"

- *Is there anything else you would like to say?*

This question is mandatory due to the fact that interview boards must give you a chance to say everything you wanted to say. This question also avoids a potential protest after the interview is concluded and is based on the concept of a potential allegation that you did not have an opportunity to say something you felt was vitally important. When you say, "Thanks, the past half hour has been a slice of paradise," you have in effect said, "I have said it all and you have given me a chance to say everything I wanted to say."

2. Interview boards are looking for candidates who understand the position for which they are applying.

Let's approach this concept from two different yet interrelated perspectives. First, assume your interview is for the position of captain. Are you prepared to tell an interview board what the position of captain is? Remember, in the previous section the word succinctly was used. Webster's dictionary defines succinctly as *clearly and briefly*, and *concise and to the point in speaking*. If you are unable to define the position you are seeking, how are you going to convince the interview board you are ready for the position if you don't know what it is? You should be able to look the interview board members in the eye and confidently say, "The position of captain is comprised of..." then briefly list the elements.

To develop a basic understanding of the position of captain (or any other position), consider the position as comprised of two parts—technical and practical. First, refer to the job announcement bulletin for the technical definition. Next, interview various captains (whom you respect) to determine what their responsibilities and daily activities consist of. This will provide a practical definition of the position of captain. This concept will be expanded in chapter 4.

Secondly, put yourself in the position for which you are interviewing. Assume you are presently a firefighter interviewing for the position of lieutenant. Take yourself out of the "locker room," and put yourself in the lieutenant's seat. It is extremely important to answer appropriate questions from the perspective of, *As *a lieutenant, I would...* This not only indicates to the interview board that you are thinking as a lieutenant, but more importantly, you are ready for the promotion.

3. Interview boards are looking for candidates who can and will do the job they are seeking.

How can a candidate convince an interview board he/she can and will do the job during a 30-minute interview? The answer, from an interview board perspective, is twofold. You need to first understand the parameters of the position.

Sound like a simplistic answer? Although it is relatively easy to determine what the position you are seeking is comprised of, most candidates amazingly "strike out" when this portion of the interview is addressed! Second, the purpose of an interview board is to select candidates that *convincingly* sell their qualifications and desire to step into a new position with effective results. This process yields a list of candidates listed in numerical order of their effectiveness in accomplishing their promotional goal. Consider the following:

- Your presentation should incorporate enthusiasm, desire, and confidence. You are enthusiastic because you have a qualified package (you) to offer your department, are ready to begin functioning in a new position, and are excited about that prospect. If your attitude and presentation style lack these qualities, reevaluate your desire and ability for that new position. Desire will be readily evident when you are sold on your qualifications, level of performance, and ability to step into that new position. Translation—you can do it!

- Confidence will naturally be apparent if you have invested time in developing and honing your public speaking skills before the interview. Confidence is directly proportional to your faith in your abilities, qualifications, *and* the amount of time and effort you invest in your interview preparation. As outlined in chapter 1, some candidates spend months preparing for a written examination, and then spend only a few weeks or days for an interview that may be worth 50% of the final score. Equal time should be devoted to preparing for the interview. Time invested and understanding your qualifications (strengths and weaknesses), understanding the position you are interviewing for (technically and practically), and developing the ability to effectively present yourself will determine the degree of desire and confidence you project at the interview. Translation—if you are unable to sell yourself or talk about your strong points to members of your family, how will you perform when you are talking to an interview board and "the pressure is on?"

- All candidates have weaknesses; some have more than others. It is not a weakness that counts. It is what you are doing to overcome the weakness. This concept is expanded in chapter 5.

- An interview board cannot rate you on what you don't say. Interview boards are not magic and are unable to read your mind. If you are unable to sell your qualifications and ability to do the job, who will? There are few guarantees in life. However, one guarantee is *you will not be given credit for what you don't say, or if you don't blow your own horn, no one else will.*

4. Interview boards are looking for candidates who will relate their qualifications.

It is a fact that every candidate competing for a promotion is the most qualified candidate on this planet. It is also a fact that every candidate has done something worth mentioning. When a candidate is in front of an interview board expounding on their numerous qualifications, what gives those qualifications any credibility? The answer to that important question is understanding and using the concepts of *relating* and *demonstrated performance*. To demonstrate, let's look at two different responses to the trait of leadership (a prime ingredient in every position within the fire service, including the position of firefighter).

First, the response statement *leadership is one of my strengths and an important trait for the position of captain* sounds good, but what gives the statement credibility to the interview board? To dramatically emphasize the inherent vagueness of this perspective, take the opportunity to listen to some practice interviews and notice how often candidates will offer numerous traits that are supposed to qualify them for the position being sought. Additionally, notice the general lack of credibility for these types of statements. Without any supporting evidence, qualification statements can be automatically placed under the heading of inference. In other words, you are inferring your statements automatically qualify you for the new position, which they do not. Giving examples of what you have done *and* learned in related areas gives your qualifications the necessary credibility!

The definition of relate (Fig. 2–4) is *to connect or join two things together*. Candidates who fail to connect or join what they have done to what they have learned will fail to sell their qualifications. Every candidate who appears before an interview board has done something and has various qualifications that are commendable. However, the candidate who can list applicable qualifications, tie those qualifications to what has been learned, *and* relate how that will help the candidate qualify for the new position *will* sell their product. Additionally, specific examples of what you have learned will demonstrate your performance capability in the particular area you are trying to justify. This combines the principle of demonstrated performance with relating and is expanded in chapter 4.

Fig. 2–4 Relate is to Connect or Join Two Things Together

Let's look at a second response to the trait of leadership and bring together the principle of relating and demonstrated performance:

Leadership is one of my strengths and an important trait for the position of captain. Prior to my career in this fire department, I commanded an infantry platoon in the U.S. Army. Since I have been a member of this department, I have had the opportunity to be an acting captain. These two specific areas of responsibility have allowed me to make decisions, analyze my actions, understand the importance of time management, training, discipline, and successfully motivate people to accomplish necessary goals.

The difference between the two preceding presentations is the first presentation operated on the principle of inference, and the second presentation operated on the principle of relating and demonstrated performance. It can also be noted that the second presentation justified the candidate's qualification of leadership.

Summary Checklist

1. The three main portions of an interview are the opening, body, and closing.

2. Interview boards are diligently looking for candidates who understand the position for which they are applying.

3. Interview boards are also diligently looking for candidates who can and will do the job they are seeking.

4. Interview boards are looking for candidates who will relate their applicable qualifications and what they have learned to the job they are seeking.

5. Always put yourself in the position for which you are interviewing.

6. The concept of demonstrated performance gives credibility to the area you are trying to sell and justify.

7. Know your weaknesses and what you are doing to overcome them. Additionally, do not discuss a weakness without turning it into a positive attribute.

8. Interview boards form an opinion of a candidate in a short period of time.

9. Interview boards must ask the following three basic questions:

   • What are your qualifications?

   • What is the job for which you are interviewing?

   • Is there anything else you would like to say?

3. Opening Impressions

Introduction

In chapters 1 and 2, we discussed the importance of enhancing your public speaking skills (for this discussion, defined as presentation skills), confidence during your interview presentation, and the basic factors comprising the makeup of an interview. One of the areas mentioned was the importance of the beginning, or the initial portion of an interview. When meeting someone for the first time, that person exhibits certain characteristics collectively combined to form an initial impression. Webster's dictionary defines initial as *having to do with or occurring at the beginning.* This definition can easily and realistically be applied to the interview environment—interview candidates will make an impression on interview board members during the initial portion of an interview. It is a fact that interview boards will form an opinion (positive or negative) of a candidate within 30 seconds to several minutes into an interview. Additionally, some opinions are formed the instant some candidates walk through the door.

During the beginning phase of an interview, there are four notable impressions each candidate will display to the interview board:

1. Application, resumé, and cover letter

2. Clothes and appearance

3. Handshake

4. Seating position

These impressions can be either positive or negative, and interestingly, a candidate can control all of these impressions. It is imperative that every candidate understand and maximize these initial impressions and utilize them to begin the interview from a position of positive strength. Consider the concept that the first few minutes of an interview are vitally important. Let's look at the initial impressions candidates can control for a positive advantage, which can last for the duration of an interview, and most importantly, have the potential to affect the outcome of an interview.

Application, Resumé, and Cover Letter

The very first contact between a candidate and an interview board is the application, resumé, and/or cover letter. Without ever having met a candidate, the application, resumé, and/or cover letter will graphically display some of your characteristics to the board members. For example, can you follow directions? Is your application neat and accurate? Does your resumé show initiative and resourcefulness? Conversely, does your application look like it was filled out at the very last minute? If you agree that initial impressions are lasting, then the importance of the look and accuracy of your application, resumé, and/or cover letter cannot be underestimated.

Application

Unbelievably, some applications look like a napkin from a fast food restaurant. White ink has been used to cover errors before retyping; the application has been typed on any typewriter that was handy (the finished product echoes that fact); the application was filled out with a pencil or with a pen running out of ink; there are empty spaces that should have been filled in; and quite often there is a misspelled word (just a single misspelled word speaks volumes about your standards). What do you think any of the preceding factors told the interview board members about you before your entrance into the room? How do you want your application to represent you? Although most civil service application forms are standard issue forms and are often a reproduction of other reproductions, the following considerations will maximize the look of your application:

- First and foremost, have your application professionally typed. There is a significant difference between the "professional" look, and the "I did it myself" look.

- Ensure the application is complete, and all requested information is correct.

- Have your application appropriately reviewed for spelling, grammar, sentence structure, etc.

- If possible, center words in the requested information boxes. For example, center your last name over or under the word "last" in the name box, etc.

- Mark an "X" in the center of the appropriate boxes.

- If possible, do not use abbreviations.

- Keep a copy of your application and be totally familiar with its contents.

- Do not turn the application in at the last moment. If you suddenly remembered that something should have been included on the application, it is now too late to include the missing information.

Resumé

As previously discussed, an application is a preprinted form that requests the applicant to fill in boxes or spaces with specific information required by a personnel/fire department. Conversely, a resumé is a piece of paper that you have complete control over the look and contents, and can be used to say anything you want to say. Consider the resumé as a *value added* addition to your application. It can be used to summarize an applicant's employment, experience, education, training achievements, and other pertinent information. However, a resumé is *also* an opportunity for a candidate to briefly outline applicable qualifications that can serve as a source of questions for interview board members. Let's restate that last sentence from several different perspectives:

- Interview boards get tired of asking the same questions and look for fresh relevant material to question.

- If an interview board is limited to a specific time frame and they ask you several questions from or about your resumé, have you (to a degree) controlled or directed the questions you are asked?

- If you are asked a question from your resumé, who is the most qualified person to answer that question? I clearly recall conducting an interview for the position of engineer. During the interview, I happened to glance at the candidate's resumé and noticed that he was the manager of a vegetable sprout farm prior to entering the fire service. My first thought was, what does this have to do with driving fire apparatus and pumping water? At the end of the interview, I asked the candidate how his employment at a "sprout farm" related to the position of engineer. He quickly smiled, looked me straight in the eye, and confidently said:

> *Chief, prior to entering the fire service, I was the manager of a sprout farm and was responsible for more than $1 million worth of product. It was my responsibility to ensure the sprouts were correctly ordered and planted, supervised, picked at the right time, shipped to the proper customer, and invoices were properly forwarded in a timely manner. As a result of this opportunity, I learned the value and importance of planning, dependability, time management, accurate records and reports, and customer service. These are the same traits I will apply to the position of engineer!*

With that response, what would you have thought? I was ready to immediately pin the engineer badge on that candidate!

Just as a resumé can be an advantage, it can also be a disadvantage if not used properly. Some candidates operate on the principle of *if a little is good, then more is better*, and tend to list every available qualification they possess in their resumé and also attach certificates from every seminar or class attended. Although this is an admirable thought, it is not a practical approach for interview board members who have time constraints between interviews, need to discuss and grade the previous candidate who appeared before them, read the next application and resumé, and review any particular questions they want to discuss before beginning an interview. Therefore, there are many ways to design a resumé that is easily read within specific time constraints, and will graphically summarize your qualifications for the new position you are seeking. Let's look at two primary considerations.

Heading. Separate your qualifications (or job attributes) into significant headings. Choose four or five headings that highlight your qualifications within the attributes of the position you are seeking. Look at the simple resumé in Figure 3–1. John Smith is interviewing for the position of captain. The five areas of qualifications that Smith wants the interview board members to notice are

1. department experience

2. fire prevention

3. leadership

4. training

5. administration

RESUMÉ
JOHN SMITH

DEPARTMENT EXPERIENCE
  * Engineer
  * Hazardous Materials Responder
  * Paramedic
  * Firefighter

FIRE PREVENTION
  * Inspector
  * Sprinkler Ordinance Technician
  * Developed Lock Box Tracking System

LEADERSHIP
  * Acting Captain
  * Muster Committee Organizer
  * Naval Officer

TRAINING
  * Department Nozzles, Hose, and Breathing Apparatus
  * Seminars
  * Probationary Firefighters
  * State University, B.A. Fire Protection Administration

ADMINISTRATION
  * Company Planning
  * Reports and Records
  * Chamber of Commerce

Fig. 3–1 Sample Resumé

Notice that these five headings separate and focus attention on each area. Additionally, each heading corresponds to the attributes of the position being sought and are the primary qualifications important to the position. This concept is expanded in chapter 4.

Related qualifications. List only those related qualifications that are important. No more than five meaningful qualifications per heading will result in a relevant resumé that can be quickly reviewed (this is a major consideration), is realistic, and stands a good chance of being digested by the board members. Let's look at the following four resumés and discuss the specifics of each one in detail. Notice that each resumé is only one page in length.

- *Resumé 1* (Fig. 3–2): This resumé was submitted by a captain interviewing for the position of battalion chief. Job attributes are neatly divided into four categories (fire suppression assignments, special assignments, specialized training, and education). Related qualifications are listed under each appropriate heading, so the resumé has a structured look. Although there is a lot of information on one page, this resumé has too much information, is not easy to read in a short period of time, has unnecessary duplication (i.e., Hollywood Area, East/Central L.A., Westwood/Bel-Air, under fire suppression assignments), and unexplained abbreviations (i.e., C.O.M.O. under specialized training).

Of these four resumés, this would be my least favorite.

Fire Captain II

FIRE SUPPRESSION ASSIGNMENTS

Battalion 5   Hollywood Area   Hi-Rise/Mercantile/Commercial
                                       High Value/Life Hazard/Brush

Battalion 7   East/Central L.A.   Heavy Industrial/Commercial
                                       Grass/Housing Projects

Battalion 9   Westwood/Bel-Air   Hi-Rise/Mercantile/High Value
                                       Residential/Brush

Battalion 10   Central S.F.Valley   Hi-Rise/Comercial/Industrial
                                       High Value/Brush/V.N. Airport

Battalion 11   Central Wilshire   Dense Residential/High Life
                                       Hazard/Hi-Rise/Commercial

Battalion 17   West S.F.Valley   High Value Residential and
                                       Commercial/Hi-Rise/Brush

SPECIAL ASSIGNMENTS/DETAILS

Training Division
        Program Director, Pre-Training Academy
        Background Investigations

Department Advocate

Acting Battalion Commander   Battalion 10

Committees
        Helicopter Airborne Fire/Rescue
        Hazardous Materials Coordinating

L.A.F.D. Program Developer/Trainer
        Airborne Engine/Squad Operations
        Hazardous Materials First Responder Training Program
        Chief Officers Seminar - Operations Plan (Book 37)
        Department Seminar - Hazardous Materials II
        Firefighters Muster Seminar - H.M. Tactics/Case Studies
        Haz-Mat Task Forces 4, 27 and 39 - Tactical Operations

Evaluator
        Division Evaluations L.A.F.D. Divisions I & III
        Earthquake Conference 85 & 86 - Multi-Agency Exercises

SPECIALIZED TRAINING

Los Angeles Fire Department Programs
        C.O.M.O., Instructor Training, Department Seminars (10)
        Strategic Chemical Attack Teams Training

Hazardous Materials Seminars/Symposiums (Total of 9)
        C.M.A. (CHEM-TREC), N.F.P.A., F.E.M.A., Rockwell
        Shell Oil, Mutual Propane, Northridge Tox Center

EDUCATION
California State University Los Angeles
        Fire Protection Administration and Technology

Fig. 3–2 Resumé 1

- *Resumé 2* (Fig. 3–3): This resume was submitted by a fire inspector interviewing for the position of captain. Similar to Resumé 1, job attributes are neatly divided into categories (department experience, fire prevention, special assignments, leadership, and education). Related qualifications are listed under each appropriate heading, so this resumé also has a structured look. However, notice there is more "white space" (less writing) and the information presented is easier to quickly digest as compared to Resumé 1. Several key comments:

  ✓ Under department experience, it is unnecessary to list the specific areas (Battalion 11, etc.) and the number of years in each area. Also, to the uninformed, what does A/O mean? Remember, eliminate the fluff and no abbreviations.

  ✓ Under fire prevention, it is also unnecessary to list where the candidate was an inspector and the length of time.

  ✓ Under special assignments, why are some words after LAFD testing programs capitalized and some are not?

  ✓ Under leadership, notice that LAFD chaplain is repeated (also under special assignments). The type of church attended should not have been specified. Although it is not supposed to matter, I think you get the point. This particular candidate was an elder in his church, so it would have been more advantageous to have listed this qualification as *church elder*. Keep this item in the back of your mind, as we will return to this specific area in chapter 4.

  ✓ Under education, the candidate should not have listed the number of units completed. In this case, it is obvious to the board that the candidate has not completed the appropriate degree(s).

Of the four resumés, this would be my third favorite.

JOHN SMITH

DEPARTMENT EXPERIENCE

*	Inspector I	West/Industrial	1 year
*	Engineer	Battalion 11	6 years
		Battalion 4	1 year
*	A/O	Battalion 11	2 years
*	Firefighter	Battalion 3	3 years
		Battalion 1	2 years

FIRE PREVENTION

*	Inspector I	West Industrial Unit	1 year
*	Battalion 11		

SPECIAL ASSIGNMENTS
* LAFD Chaplain
* LAFD Testing Programs: nozzles, hose, and Breathing Apparatus
* Apparatus Maintenance Committee
* Peer Group Instructor
* Co-author of Seagrave Aerial Logbook
* Foam Carrier 34 - Training members in all divisions

LEADERSHIP
* Rookie training of basic engine and truck company operations
* Engineer and A/O candidate training
* LAFD Chaplain
* Commissioner of LAFD Racquetball
* Firefighters for Christ
* Grace Community Church

EDUCATION
* Fire Science and Business Management - 51 units
* Graduate level studies in Theology - 105 units

Fig. 3–3 Resumé 2

- *Resumé 3* (Fig. 3–4): This resumé was submitted by a firefighter interviewing for the position of captain. This resumé is superior in its presentation value to Resumés 1 and 2. However, let's analyze the following comments:

  ✓ This resumé is easy to read; job attributes are neatly divided into four areas with related qualifications under each appropriate area.

  ✓ At the bottom of the resumé, the candidate has included a short statement about goals and ideals in a box that is readily apparent. This is an excellent addition and definitely grabs your attention.

  ✓ Position sought: captain. Remember in chapter 2 we discussed the concept of putting yourself in the position for which you are interviewing? This heading does not accomplish that goal. Additionally, it tends to blend with the other headings.

  ✓ Some qualifications are followed by "x 3," "x 2," etc. Although that probably means the subject was completed two or three times, it would be better to delete these notations as they are repetitious and may be confusing.

  ✓ As a side note, the smallest print mandates the reader must look very closely to read the information (i.e., "minor in Counseling," etc.).

  ✓ An excellent question from this resumé would be to ask the candidate why he took chief officer classes for the position of captain (listed under the heading of education and training). If the candidate could not answer this question, why put that type of information on a resumé? Additionally, after the candidate delineated that he took public speaking and communications (also listed under education and training), I would hope his interview supports that fact! Whatever you put on a resumé, expect a potential question (which is the primary reason you put it there) and be prepared to support your claims.

Although this resumé is very good and is often chosen as the best resumé from the four, it would be my second favorite.

JOHN SMITH

Position Sought: Fire Captain

Background & Related Experience

- **1984 - Present   Folsom City Fire Department**

    Engine Company Officer

    MSA Repairs

    Strike Team Engine Officer x 3

- **1983 - 1984   Ramona Fire Department**

    Wildland Fires and Initial Training

Education & Training

- Fire Officer - State Certificate
- Complete Fire Officer - Paul Stein and John Mittendorf Seminar
- B.A. in Ministerial Studies

    Minor in Counseling

    Organization Behavior - Group Dynamics

    Public Speaking and Communication

- Leadership Seminars and Weekly Training
- Sacramento High-rise Training x 2
- Chief Officer Classes x 3

Administration

- Records and Reports
- Strive to cooperate and maintain effective communication with other Department Officers
- Motivated and Enthusiastic

Fire Prevention

- Customer Relations - People Person
- Public Education

I believe the ideals that I have been developing as a leader will only serve to benefit the Folsom City Fire Department's goals and mission statement. I believe that integrity, loyalty, accountability and professionalism have been at the top of my list in preparing for this position and are at the heart of my character. I strive to be a coach, role model and positive influence for those I have the privilege to work with.

Fig. 3–4 Resumé 3

- *Resumé 4* (Fig. 3–5): This resumé was also submitted by a firefighter interviewing for the position of captain. Interestingly, when students in a workshop setting are given all four of these resumés and asked to pick their favorite, none choose Resumé 1 or 2, some will choose Resumé 3, but most will choose this one. Why? Let's take a closer look:

✓ The candidate paid $15 to a resumé service for the finished product. Although there are various ways to outline a resumé, this particular resumé was designed to be simple and easily read. Additionally, the border is a nice touch and sets it off.

✓ Again, job attributes are neatly divided into areas with related qualifications under each appropriate area.

However, let's slightly fine tune this resumé.

✓ Delete the position sought: fire captain (remember the principle of assuming the position for which you are interviewing).

✓ Delete all of the dates under special assignments (remember to delete the fluff).

✓ I would question the 1976–1986 build and supervise new home construction projects under background and related experience. This indicates the candidate was "working on the side" during his employment with the San Diego City Fire Department. A civilian interviewer might not appreciate that fact. In fact, it would also be a good idea to drop the dates under background and related experience.

✓ As an interesting side note, the enticement of this resumé *might* be increased if a few qualifications under special assignments, education–training, and commendations and recognitions were deleted and a similar box inserted (bottom of page of Resumé 3) that included a short statement about goals and ideals.

✓ Although I tend to stay away from numbers, the numbers preceding the items under commendations and recognitions clearly indicate this candidate is way above the norm in this area.

JOHN SMITH

Position Sought: Fire Captain

Background & Related Experience
- 1974 — Present   San Diego City Fire Department
- 1976 — 1986   Build & supervise new home construction projects
- 1970 — 1974   Sergeant United States Air Force

Special Assignments
- 1982 Hose Repair Department
- 1981 Wildland Task Force
- 1978 - 1979 Training Department
- 1974 - 1976 Fire Department Communication Center
- 1988 Training Library Committee
- 1983 Organizational Development Program

Education — Training
- A.S. Degree Fire Science
- Fire Officer Instructor Series Completed
- Affirmative Action Equal Employment Opportunities Seminar
- Swift Water Rescue
- Strike Team Leader
- Fire Arson Detection
- Supervisors Effectiveness Course

Commendations & Recognitions
- 1 Fire Department Life Saving Commendation
- 4 Special Assignment Commendations
- 1 Outstanding Fire Fighter Commendation
- 2 Community Life Saving Commendations
- 1 Community Volunteer Commendation
- 6 Fire Department Letters of Recognition

Fig. 3–5 Resumé 4

Resumé tips.

- Before preparing a resumé, determine if it will be accepted for your interview. Also, if a resumé will be accepted, can it be given to the receptionist prior to your interview? If possible, give your resumé to the receptionist when you arrive. If you turn it in with your application, it may get lost, but it probably *will* have wrinkles, fingerprints, smudges, and other distractions on it. If you give it to the receptionist, it stays the way you want it—perfect!

- Use quality paper, but don't overdo it and don't use scented paper. This is a serious interview, not a cosmetic counter at a department store.

- Have your resumé professionally typed and/or completed. There are numerous "resumé services" available that can prepare an excellent resumé for a minimal cost.

- Provide a copy for each interview board member.

- Separate the job attributes (or your primary qualifications) into headings that address the elements of the position being sought. Next, list your related qualifications under each heading. Limit your related qualifications to no more than five items.

- Eliminate "stuff and fluff." Items such as your marital status, address, phone number, age, etc. are probably on your application and are definitely not primary job qualifications.

- Make the resumé easy to read and summarize as much as possible; consider the words brevity and succinct.

- Limit the resumé to one page. The longer you make your resumé, the more you increase the chance it will not be beneficial.

- Know the contents of your resumé and be prepared to expand on each heading and related qualification.

- Keep a copy of your resumé if it needs to be attached to your application when submitted.

Cover letter

A cover letter can be used as a letter of introduction from you to the interview board and is submitted in addition to and with your resumé. Sample one-page cover letters are illustrated in Figures 3–6 through 3–8. Let's look at and discuss the specifics of each one:

- *Sample cover letter 1* (Fig. 3–6): Although this cover letter appears to be one page, the typing has been double spaced to create "white space." This enhances the look and the ability to read and digest the information *quickly*. Of primary interest, the letter quickly overviews the candidate's background and states he is looking forward to meeting the interview board members.

COVER LETTER

DATE

Interview Board Members
Somewhere Fire Department
111 Main Street
Somewhere, CA 91102

Board Members:

Let me start with an introduction. My name is John Smith. I
have been a member of the Somewhere Fire Department for nine
years. My experience varies widely from engine and truck
assignments to the Fire Prevention Bureau. Additionally, I
have served as the President of the Firefighter's Union and
on various fire department committees.

As my resumé indicates, my experience and formal education
makes me a highly qualified person to be considered for this
promotion.

I am looking forward to our meeting and will thank you in
advance for taking the time to interview me.

Sincerely,

John Smith

Fig. 3–6 Sample Cover Letter 1

- *Sample cover letter 2* (Fig. 3–7): This cover letter is my least favorite for three basic reasons:

  ✓ The letter cannot be easily read and digested in a short period of time.

  ✓ Most importantly, it probably repeats the same information on the candidate's resumé. If a cover letter is repetitious or similar to a resumé, then either the resumé or cover letter should be deleted.

  ✓ Notice the challenge to the interview board members in the third paragraph. If this candidate is not a good communicator during the interview, or does not perform efficiently when given several *stressful* questions, then how credible is this cover letter?

COVER LETTER

DATE

Interview Board Members
Somewhere Fire Department
111 Main Street
Somewhere, CA 91102

Dear Board Members:

I would appreciate a few minutes of your time to read the enclosed resume which summarizes my qualifications for advancement to the Lieutenant's position with the Somewhere Fire Department. I am seeking this position where my extensive training, experience, and abilities would have valuable applications.

My background includes seven years of fire service experience which includes six years working in all different areas of the city, and one year in the Fire Prevention Bureau. I am a state certified Fire Officer and have an Associate of Science Degree in fire technology. I have served as the president of the Firefighter's Union and have been a member of the department's Uniform and Training Committee.

I am a self-motivated, dedicated person who is able to coordinate and communicate well with management and personnel at all levels. I function very efficiently within decision making environments under pressure and in rapidly changing situations, and consistently produce measurable results.

Since only a small portion of all of the elements of my experience and capabilities can be presented, or assessed in this letter, I am sincerely looking forward to meeting and talking with you during out time together.

Sincerely,

John Smith

Fig. 3–7 Sample Cover Letter 2

- *Sample cover letter 3* (Fig. 3–8): This is my favorite cover letter compared to the other two. It is easy to read and digest, is presented well, does not repeat Smith's resumé, and talks about the ability of the candidate to make a positive influence on his department and the personnel he has had the privilege to work with.

COVER LETTER

Date

Interview Board Members
Somewhere Fire Department
111 Main Street
Somewhere, CA 91102

Dear Board Members:

My name is John Smith and I am interviewing for the position of Engineer. I believe my character which includes integrity, loyalty, accountability, and professionalism, are at the heart of what make me desirable for the position of Engineer with the Somewhere Fire Department.

I have always taken pride in the professionalism of this department and its progressive position in the fire service.

As an Engineer, I will be able to make a greater impact in the maintenance of our professional image and in helping the department remain on the leading edge of the fire service.

I strive to be a role model and a positive influence for those that I have had the privilege to work with. My desire to expand my capabilities as a coach and teacher will benefit me as well as the Somewhere Fire Department.

Sincerely,

John Smith

Fig. 3–8 Sample Cover Letter 3

Cover letter tips.

- If for some reason it is necessary to choose between a cover letter and a resumé, choose the resumé.

- If the interview board members are from your department and know you, a cover letter could be repetitious. Conversely, if the board members are from outside departments and/or civilians, a cover letter can be very beneficial.

- A cover letter is another piece of paper to be read within specific time constraints, so keep the contents brief. Remember the KISS principle: Keep It Simple Stupid.

- Provide a copy for each interview board member.

Your paperwork (application, resumé, cover letter) is their *first* impression of you.

Clothes and Appearance

When you first enter the interview room, the interview board members will immediately notice how you present yourself. Specifically, are your clothes (uniform or civilian) new or average in appearance? Do your shoes look like a mirror or like you just completed a 10-mile hike? Are your grooming standards neat or in need of attention (Fig. 3–9)? The way you look and dress will initially put you with the pack or ahead of it. The choice and opportunity is yours.

Fig. 3–9 Look Your Best

Clothes and appearance tips.

- If you have a choice between your department dress uniform and civilian clothing, I recommend your department uniform. It graphically illustrates your pride in your department, has a specific professional look, and does wonders for your pride and confidence. For example, what is your initial thought when you see a U.S. Marine in dress uniform?

- If you wear civilian clothing, you can go for the "power look" if it makes you feel better. It won't hurt but will probably not affect your grade.

- If you are serious about your interview, purchase a new uniform or suit (unless it is already new) and have it tailored. Tailoring is a minimal cost expense but will make clothing fit properly and look its best. None of your competitors will look better (or feel better) than you do, and the interview board members will notice.

- Rent a pair of leather patent shoes (the ones that look like a mirror). When you see someone in military dress uniform, do you notice their "spit-shined shoes?" Absolutely! Will the interview board members notice yours? Absolutely! I recall an interview in which a candidate used this concept to perfection. This particular department required each candidate to wear the dress uniform, including the hat. When I asked the candidate to sit down and get comfortable, the candidate placed the hat at the end of the table, and in plain view of the board members.

During the interview, I noticed the hat appeared to be brand new (they normally look like they have recently been discovered by a paleoarchaeologist). At the end of the interview, I asked the candidate if the hat were new. Looking straight into my eyes, this was the exact comment, *Chief, my hat, uniform, shoes, socks, and undergarments are all brand-new. I am prepared and ready for this new position.* Not surprisingly, he was absolutely right.

- Pay particular attention to your grooming. Get a haircut about a week before your interview. This will allow your hair to fill in slightly and allow the skin around your hair to have an even color.

- Minimize visible jewelry, ornaments, tattoos, and heavy scents. You don't need to look like a Christmas tree and/or run the risk of alienating a board member. Stay neutral.

This is their *second* impression of you.

Handshake

Prior to your interview, practice shaking hands and looking people *in the eye* when you meet them. You want to develop a firm and positive handshake coupled with the ability to look the members of the interview board in the eye when you are introduced. A "cold fish" handshake does little to enhance your image and will be remembered during the duration of your interview.

Handshake tips.

- If possible, make a concentrated effort to remember the name of each board member, and if possible, address each member by name during and at the end of the interview.

- If you sweat under pressure, quickly dry and warm the hand you will use to greet the board members. Just before entering the room, quickly and forcibly wipe your hand on the side/rear of your pants. This will remove moisture and restore some warmth to your skin.

- When you practice shaking hands, keep your eye on the other person's hand until both hands meet. If you play golf or racquetball, you are aware of one of the most important rules of the game—*keep your eye on the ball.*

- If you have recently been lifting weights at a local gym, it is not necessary to let each board member know you are capable of crushing his/her hand with a single squeeze (Fig. 3–10). A firm positive handshake is sufficient.

This is their *third* impression of you.

Fig. 3–10 A Firm Positive Handshake is Not Like Lifting Weights at a Local Gym

Seating Position

After meeting the members of the interview board, they will tell you to have a seat on the chair provided for that purpose. Usually, the chair is located in close proximity to the interviewer's table. Depending on the location of the chair, consider moving it so it is about four or five feet back from the table. This will accomplish the following:

- You will have positioned yourself for better eye-to-eye contact with the members of the interview board. The closer you are to multiple individuals, the more difficult it is to maintain *natural* eye-to-eye contact without your head resembling a rotating beacon (Fig. 3–11). During the interview, concentrate on maintaining eye contact with each board member and do not primarily focus on the board member who asked you a question.

- You will have signaled to the interview board *psychologically* that this is your interview, this is where *you* want the chair, and you are *ready* to go.

This is their *fourth* impression of you.

Fig. 3–11 Seating Position Can Be Important

At this point, the interview board is ready to initiate the verbal question-and-answer portion of the interview. However, evaluate the preceding four considerations we have discussed, and analyze the initial impressions you can make before the question portion of the interview begins.

Summary Checklist

1. Interview candidates quickly form an impression on interview board members during the initial portion of an interview.

2. Candidates can control four initial impressions that occur during the initial phase of an interview.

3. The first impression of a candidate on an interview board is the application, resumé, and cover letter.

4. A resumé is a summary of an applicant's employment, experience, education, and other pertinent information.

5. Resumé qualifications should be important and separated into significant or major headings that are followed by related qualifications.

6. A cover letter is a letter of introduction from you to the interview board and can be submitted in addition to and with your resumé.

7. The second impression of the candidate on the interview board is the candidate's clothes and personal appearance.

8. Clothes and personal appearance will put the candidate with the pack or ahead of it.

9. The third impression on the interview board is the handshake, along with the ability to look each board member in the eye, and if possible, remember each member's name.

10. The fourth impression on the interview board is the seating position of the candidate, along with the ability to maintain good eye-to-eye contact with each board member during the interview. Natural eye-to-eye contact with each board member during the interview is vitally important.

4. Preparing a Response to the Opening Question

Introduction

Once you have been seated on the chair near the interview table, it is time to begin the next portion of the interview—the opening. Normally, the board will attempt to put you at ease (if that is possible with a blood pressure of 340/270, a mouth full of cotton, and sweat cascading down your face), and try to have you talk by asking you a simple introductory question. The opening or introductory questions are normally similar to the following:

- What are your qualifications?

- What is your background?

- Why do you want the job?

- Where have you worked?

- What is the position for which you are interviewing?

- Can you give an overview of yourself?

If you have had the opportunity to take a prior interview, do these questions sound familiar? Although the preceding questions are similar, straight forward, and are commonly utilized to start a promotional interview, they subtly accomplish three unique objectives:

1. Help a candidate relax by talking about a subject the candidate is most familiar with. Give a candidate an opportunity to discuss something *about* himself/herself.

2. The interview board members have given the candidate an opportunity to talk about two items that must be discussed during an interview: (1) What is the position for which you are interviewing? and (2) What are your qualifications for the position?

3. Set the standard for the balance of the interview. Either you are prepared and start on a confident high note, or you flounder around and try to improve during the interview (which is not highly recommended or easily accomplished).

As emphasized in chapter 3 and as you should now be aware, there are four impressions you will make on the interview board members before you are asked your first question. As a result, you should be fully prepared to take advantage of those initial impressions. Similarly, with prior knowledge of what general type of question you can expect during the opening portion of an interview, you should also be totally prepared (*prepared to the point that you are confident, enthusiastic, and ready*) to answer the question asked. Unfortunately, this is the portion of an interview where most candidates either strike out (Fig. 4–1) or do not utilize a golden opportunity to its full potential. Although a simple straight forward answer to any of the opening questions is commonly utilized by 99% of interview candidates, this same percentage of interview candidates will also either fail or ineffectively incorporate the following considerations with their answer:

- Overview the new position for which they are interviewing.

- Overview their qualifications for the new position.

- Most importantly, relate their qualifications to the position for which they are applying.

Fig. 4–1 Don't Miss the Principle of Relate

Although the concept we have been considering may sound simple, few candidates understand and utilize this concept. For example, let's consider a response to the opening question, "What can you tell us about yourself?" An abbreviated typical answer to this question is as follows:

After graduating from high school, I attended City College for two years. I graduated with a major in business and then worked for several years for a marketing company while simultaneously serving in the Air National Guard. I was then appointed to this fire department where I have been a member for the past 10 years. During that time, I have been a firefighter-paramedic for 6 years, and 4 years ago I was appointed to the position of engineer. As an engineer, I have had the opportunity to chair the department's apparatus and equipment committee who is responsible for developing specifications and purchasing fire apparatus and appropriate equipment. Currently, I am in the process of testing for the position of lieutenant for this department.

If you feel the candidate answered the question and your overall opinion of the response is positive, you are partially correct. The positive aspect of this answer consists of the following considerations:

- The candidate answered the question.

- The candidate has a solid background consisting of education, good work background, promotion within the department, and responsibility (apparatus and equipment committee).

However, the answer failed to tie the candidate's background (which is excellent) to the position of lieutenant. Failure to tie or connect a background and any appropriate qualifications to the position being sought is assuming an interview board will *infer* that a background and/or qualifications will automatically qualify a candidate for a promotion (which they do not). As a side note, Webster's dictionary defines infer as *to conclude by reasoning from something assumed; to imply.* Seriously consider the fact that *all* candidates have backgrounds and qualifications that can be used to answer the preceding question (or any typical opening type of question). The real question is, "What has set you and your background and qualifications apart from the rest of the candidates?" The answer is (as applied to the preceding question) quickly overviewing your background and qualifications while simultaneously *relating* them to the position being sought. Sound simple? Although it is, few candidates utilize this concept and wonder why their final interview score is not higher.

Organizing Your Opening Response

To begin, let's consider a method you can use to answer your opening question while *simultaneously* accomplishing the following goals:

- Be able to confidently and effectively start your interview.

- Answer the question that has been asked.

- Overview the position you are seeking with your qualifications.

- Relate your qualifications to the position.

- Begin to set the standard for the balance of your interview.

Remember, opinions are formed in a short period of time and if there is an opportunity to place yourself "ahead of the pack," use it. Additionally, let's use every available opportunity to enhance these five goals. Based on the concept that an interview board will ask you a typical opening question similar to the questions we have considered at the beginning of this chapter, let's consider how to put your opening answer together.

First, before you ever sit down on the "chair" and answer the first question of your interview, you must spend a noteworthy amount of time in preparing your response to any typical opening type questions (unless you are comfortable in "winging it"). How do you do that? If you take a class in English or report writing, you are always advised to initially make an outline of your subject before writing a report. Applying the same principle to our discussion, you should first make an outline of your responses to typical opening questions. The outline can then be expanded and *fine-tuned* to yield your desired finished product. Additionally, the main points of an outline can be easily reviewed in your mind while responding to a question and also minimize forgetting key points.

Secondly, if the interview board must ask you about the position for which you are applying *and* your qualifications for that position during the interview, take some initiative and incorporate the concept of answering the questions that have been asked while simultaneously overviewing the position with the qualifications for that position. To accomplish this seemingly difficult task, consider a method to develop an opening response (which can be a difficult portion of an interview), and ensure that you overview and *relate* your qualifications while answering the opening type question. As a starting point, consider the most commonly promoted positions in the fire service:

1. Engineer

2. Sergeant

3. Lieutenant

4. Captain

5. Battalion chief

Applying the principles we are currently discussing, divide these positions into three categories: (1) engineer, (2) sergeant, lieutenant, captain, and (3) battalion chief. Now, in order to keep things simple, what is the primary difference between engineer and the other positions? The answer is supervisory responsibility. Next, what is the primary difference between sergeant, lieutenant, and captain? The answer is span of control (supervision is a common denominator to these three positions). Finally, although span of control is a common denominator to the positions of sergeant, lieutenant, captain, and battalion chief, what is the primary difference between the position of battalion chief and the positions of sergeant, lieutenant, and captain? The answer is the primary focus of a battalion chief is management.

Utilizing these basic guidelines, let's consider these positions and how to quickly develop an outline focusing on the primary job elements of the position you are seeking in relation to your qualifications. Any position can be considered multi-dimensional (Fig. 4–2). The horizontal dimension consists of the different elements of that position, and the vertical dimension consists of your qualifications within or under the horizontal elements. First, let's apply the horizontal and vertical elements to the position of engineer, and then we will consider the other positions.

Fig. 4–2 Any Position Can Be Considered Multi-Dimensional

Engineer

Initially, the position of engineer can be conveniently divided into five job elements (which will be the horizontal dimension) as follows:

1. **Department experience.** Consists of actions necessary to mitigate all types of emergency and non-emergency incidents.

2. **Fire prevention.** Consists of occupancy inspections, pre-fire planning activities, and equitable enforcement of applicable laws and codes.

3. **Apparatus operation/maintenance.** Consists of apparatus maintenance, operation at incidents, driving, and other related responsibilities.

4. **Leadership/training.** Consists of *limited* opportunities (under the direction of a company officer) of instructing company firefighters in proper apparatus maintenance, operation, driving, and other similar responsibilities associated with apparatus.

5. **Customer service/public relations.** Can be utilized to overview your ability to represent your department in a positive and professional manner, on and off duty.

Notice that these five areas encompass the main areas of responsibility for the position of engineer. Obviously, these areas and their related contents will vary depending on your opinion. However, they are your opinion and they also form an outline and a starting point that can be expanded to overview any position or job. Additionally, these areas are easy to remember in an interview environment, and also happen to be similar to the headings on the resumés in chapter 3. This approach will result in the following benefits for an interview candidate:

- Dividing a position into main areas of responsibility that defines the position and is easy to remember.

- Keeping the information you must remember simple and consistent, yet effective.

- Your resumé will mirror the areas you feel are important for the position you are seeking and may entice the board to ask you a question about the contents of your resumé.

Next, list all of your qualifications under each area similar to the following example:

- **Department experience.**

  ✓ engine and truck company experience

  ✓ haz mat responder

  ✓ certified paramedic

  ✓ Incident Management System (IMS) experience

- **Fire prevention.**

  ✓ station fire prevention coordinator

  ✓ records and inspection forms

  ✓ sprinkler ordinance program

- **Apparatus operation/maintenance.**

  ✓ automobile repair mechanic

  ✓ department driver training program

  ✓ acting engineer

- **Leadership/training.**

  ✓ conducted company training programs

  ✓ chairperson apparatus and equipment committee

  ✓ Associate of Arts degree in fire science

- **Customer service/public relations.**

  ✓ Little League coach

  ✓ Fire Service Day coordinator

Any qualification that you feel is relevant should initially be listed under each appropriate area. This takes work and serious thought. When you have completed your outline, overview the information that you have developed. Now, throw out the items that are not primary qualifications, and items you are not totally satisfied with (eliminate the fluff). Remember, this is not the "Gong Show." With additional thought on your part, take several weeks to develop four or five *noteworthy* items under each area that are relevant, truthful, and you are comfortable with. If you don't believe it, nobody else will!

Your finished multi-dimensional overview should look similar to the preceding example. With this process, you have now developed an outline of the areas of responsibility for the position of engineer and your qualifications for the different areas for that position. With additional thought, you should now be able to define the position of engineer, define and elaborate on your qualifications for the different areas of responsibility, and be able to *relate* those qualifications to what you have learned and how that will assist you in accomplishing the job of engineer.

Now, let's consider the common opening interview question, "What is your background?" and consider two different approaches to answer this question. Response 1 (Fig. 4–3) utilizes the multi-dimensional overview.

Sample Response #1 for Engineer

Members of the board, I have 11 years of experience with this department. This experience includes the positions of firefighter, paramedic, and hazardous materials responder. I have worked in the high-incident areas of the city and have been exposed to a wide variety of emergency incidents.

My fire prevention background includes experience as a station fire prevention coordinator, and I have familiarized myself with the records and reports that are required for the maintenance, operation, and necessary compliance with this program. Additionally, I have been involved in the city's new sprinkler ordinance program.

My background in the area of apparatus operation and maintenance includes extensive experience as an automotive repair mechanic and a driving instructor in the department's driver training program. I have also been an acting engineer during the past three years, operating engine apparatus at several large emergencies.

My leadership experience consists of acting as the chairperson of our muster committee for the past two years. My training qualifications consist of researching and developing a training program for the department's current nozzles, hose, and breathing apparatus. I have conducted several building construction seminars for this department and have assisted in training probationary firefighters. My education consists of an Associate of Arts degree in fire science from a local college.

Finally, my community involvement consists of coordinating this department's Fire Service Day program, and I am a coach for a Little League team.

Fig. 4–3 Sample Response 1

Notice this response answered the question, "What is your background?" and overviewed an excellent background with numerous qualifications. However, this response also required that the interview board must automatically assume this overview qualified the candidate for the position of engineer through the process of inference. Additionally, this response never specifically related any (background) qualifications to the position of engineer. Response 2 modifies this presentation for the second approach to the same question (Fig. 4–4).

Note the difference. Response 1 summarized the candidate's background (which is likely to be similar to other candidate's responses to the same question). However, it failed to tie any of the candidate's background considerations to the position of engineer. Response 2 also summarized the candidate's background, but simultaneously incorporated the following considerations:

- The first two sentences were utilized to answer the question, "What is your background?" Then the candidate asked permission to answer two necessary questions (what is the position for which you are interviewing, and what are your qualifications). Unless the interview board changes the subject, they have given you the green light to "go." It is very doubtful that an interview board would not want to hear what comprises the position of engineer in relation to your qualifications.

  Further, you are now prepared to tell the interview board, in your own style, what the position of engineer is, how you are qualified in each area of responsibility, *and* relate what you have learned that will qualify and assist you in the position of engineer.

- The position of engineer was then described (the five areas of responsibility), and each of the areas were delineated with appropriate qualifications while simultaneously *relating* how these qualifications apply to the position.

Sample Response #2 for Engineer

Members of the board, I have had the opportunity to be a member of this department for 11 years. I have been assigned to engine and truck companies in high-incident areas and I am currently certified as a paramedic and hazardous materials responder. However, I am currently applying for the position of engineer. Let me take a few minutes and describe the position of engineer, and how I am qualified to be an effective engineer in this department.

In my opinion, the position of engineer is comprised of five areas of responsibility. They are experience, fire prevention, apparatus operation/maintenance, leadership/training, and customer service/public relations.

In the area of experience, I have evaluated and applied the various assignments and opportunities the department has made available to me. I have 11 years of varied experience that is noteworthy from the perspective of accumulated experience, a combination of various assignments, and the time I have left to give this experience back to this department. My field assignments have been in areas classified as high-incident, commercial, residential, and I am a certified paramedic and haz mat responder. I have utilized these assignments to establish and expand my knowledge of fireground operations from a strategic and tactical perspective, and as an engineer, develop guidelines to deploy apparatus necessary to mitigate various incidents and consider other types of apparatus placement needs.

Fire prevention is a necessary element within the position of engineer, which affords the opportunity to gain experience as a station fire prevention coordinator and become familiar with the records and reports that are required for the maintenance and operation of the program. This experience will ensure that as an engineer, I will be able to assist my company officer in the timely and necessary compliance required during routine inspections in concert with the completion of required records and reports. My involvement in the city's new sprinkler

Fig. 4–4 Sample Response 2

ordinance program has expanded my background in specialized code requirements involving specific company inspections that as an engineer, I will be able to assist other company members.

The operation and maintenance of fire apparatus from the perspective of emergency and nonemergency operations is a cornerstone of the position of engineer. With an extensive background as an automotive repair mechanic, I am familiar with automotive theory and diagnostic methods that will assist me as an engineer to properly maintain my apparatus. As a driving instructor in the department's driver training program, I understand the importance of proper driving techniques and will be able to use this knowledge in properly and safely operating fire apparatus from the perspective of routine and emergency operations, and the way my driving is perceived by the citizens in our community.

The opportunity to serve as acting engineer at several large emergencies has allowed me to consider the proper spotting of apparatus, both from the perspective of the initial company and later arriving companies and how these considerations can enhance an incident commander's options and priorities. Acting engineer has also required that I am familiar with the operation of apparatus and fire service water hydraulics and aerial device theory. This experience will enhance my ability as an engineer to operate either engine or truck apparatus.

Although an engineer is primarily concerned with the operation and maintenance responsibilities of fire apparatus, leadership can be an important portion within the position of engineer. The opportunity to act as an engineer has underscored my ability to get a job done by providing necessary accountability in the absence of the regular engineer. As chairperson of our muster committee, I have utilized short and long range plans, organized necessary resources, directed and controlled these resources, provided clear and effective communications, and ensured accurate and timely reports. Each of these areas contains the same tools I will use as an engineer to ensure my effectiveness within my company responsibilities.

Fig. 4–4 Continued

Within the area of training, I have been able to research and develop a training program for the department's current use of nozzles, hose, and breathing apparatus. This has expanded my knowledge, utilization, and ability to maintain these items, which is an integral portion of the position of engineer. Additionally, I have been able to conduct some building construction workshops for this department and have assisted in training probationary firefighters. This has expanded by ability to deliver relevant training programs that will assist me in helping to train company firefighters in the proper operation and maintenance of fire apparatus. My degree in fire science will give me a technical background to draw from as an engineer.

Finally, as a Little League coach and a coordinator for the department's Fire Service Day program, I realize the importance of representing this department in a professional and positive manner to the citizens of this community. As an engineer, I will be able to use this responsibility when I am in the public's eye, specifically when driving my apparatus and conducting fire prevention inspections.

Members of the board, each of these areas contains the same tools and qualifications I will use as an engineer within this department.

Fig. 4–4 *Continued*

- The response utilized the words *opportunity*, *familiar*, *assist*, *made available*, etc. These words denote a sense of humility and thankfulness as opposed to some candidates who give the impression *they have done it all*. Fact: there is a major difference between these two perceptions.

- The phrase *as an engineer* has placed the candidate *in* the position of engineer. Some candidates place themselves in their current position, and not in the position they are interviewing for when answering a question. When possible, place yourself in the position for which you are applying. This perspective tells the interview board you are comfortable in the new position, *just officially promote me and pay me for it!"*

The first response used the word *engineer* once. The second response used the word 22 times. Although this might be overkill, I think you can see the difference between these two applications (the candidate in the second response is thinking as an engineer, the candidate in the first response is thinking as a firefighter).

- The phrase *I have 11 years of varied experience, which is noteworthy from the perspective of accumulated experience, and the time I have left to give this experience back to this department* is important:

  ✓ If you have minimal time and experience in your department (i.e., four to seven years), you have the opportunity to give a significant amount of your time, abilities, and qualifications back to your department. This is a method to minimize your perceived lack of time and/or experience.

✓ Conversely, if you have a noteworthy amount of time in your department (i.e., 18+ years), then you can offer the interview board a considerable amount of experience in the position for which you are interviewing. This is a way to minimize the perception of *you only have a few years left.*

- Be careful with education. Education is not the foundation or primary qualification for a new position. It only has the ability to provide you with the technical background to assist you in your new position and does not replace practical experience. If you have a noteworthy amount of education, do not alienate a board member (who might have minimal or no education) as you eloquently describe your extensive educational background. Balanced qualifications are a combination of education and practical experience.

- Under the area of experience, the candidate used the terms *strategical* and *tactical* to refer to a specific implied knowledge of fireground operations. If you use *trick or snappy* terms/ phraseology, be able to define them. For example, an excellent question for the interview board to ask this candidate is, "What is the definition of strategy and tactics as applied to an engineer?" Additionally, under the area of leadership, the use of short and long range plans was briefly mentioned. This candidate should also be prepared to delineate this concept and/or describe the short- and long-range plans that were formed and used. Always consider the KISS principle.

- Last but not least, the second response concluded with the candidate's knowledge of the importance of representing the candidate's fire department in a professional and positive manner (on and off duty) to the citizens of the community. If you use this concept in your interview, don't be surprised if the board members hit the floor in amazement and reward you with an outstanding score (on this issue).

Remember, during the length of your interview, you are your own salesperson. You cannot be evaluated and graded on what you do not say!

Sergeant, lieutenant, and captain

Now, let's apply the horizontal and vertical elements to the positions of sergeant, lieutenant, and captain. As previously delineated, these positions share a common trait—supervision. However, there is an increasing span of control and responsibility when promoting up through these ranks. From this perspective and similar to the position of engineer, these positions can also be conveniently divided into five basic job elements (the horizontal dimension):

1. **Department experience/operation.** Consists of actions necessary to mitigate all types of emergency and non-emergency incidents.

2. **Fire prevention.** Consists of occupancy inspections, pre-fire planning activities, and equitable enforcement of applicable laws and codes.

3. **Leadership.** Consists of the ability to direct the people under your command/supervision.

4. Training. Consists of developing and implementing necessary and applicable training programs for personnel under your command/supervision. It can also include your educational achievements.

5. Administrative. Can be utilized as a "catch all," and can include items such as records, reports, goals, routine operations, and most importantly, customer service/public relations.

Interestingly, these five areas encompass the main responsibilities for the positions of sergeant, lieutenant, and captain. Obviously, these areas and their related contents will vary depending on the specific job description and your opinion. However, they are your opinion and they also begin to form an outline and a starting point for these positions. Additionally, these areas of responsibility are easy to remember in an interview environment, and are similar to the headings in the resumés in chapter 3. Next, list all of your qualifications under each area similar to the following example (the vertical dimension):

- **Operations**

  ✓ Bachelor of Arts and Associate of Arts degree

  ✓ acting captain

  ✓ strong fireground experience

  ✓ emergency medical technician (EMT) and member haz mat, high angle rescue, confined space rescue, and urban strike teams

- **Fire prevention**

  ✓ certified fire code inspector

  ✓ conducted presentations to Rotary and Lions clubs

- **Leadership**

  ✓ assistant manager for retail outlet

  ✓ forestry engine foreman

- **Training**

  ✓ developed dispatcher training program

  ✓ CPR instructor–trainer

  ✓ developed and implemented promotional training programs for firefighters

- **Administrative**

  ✓ developed annual reporting system

  ✓ developed current records management system

Now, consider the common opening interview question, "What are your qualifications?" Utilizing the preceding principles, consider an answer to this question for the position of captain. Response 3 utilizes the multidimensional overview (Fig. 4–5).

Sample Response for Captain

Members of the board, I have been with this department for 10 years. I have promoted through the ranks to my present position of lieutenant. While pursuing promotional opportunities, I have also used every opportunity to expand my knowledge base, my skill level, and my worth to this fire department. I am applying for captain so I may utilize my education and experience to make a greater impact on this department. Let me explain the five major areas of responsibility for captain and why I am your best-qualified candidate.

The position of captain has five major areas of responsibility: operations, training, fire prevention, leadership, and administration.

In operations, my educational background gives me the theoretical knowledge to apply and combine my practical experience. I have a Bachelor of Arts degree from Acme University, a Fire Science degree with honors from Smith College, and have completed dozens of fire service seminars and classes. Seven of my ten years have been spent at our busiest station, which has provided me with a strong practical experience base. I have been an acting captain for the past five years. My fireground and emergency operations skills have been honed over this time and I have shown I can evaluate varied emergency situations and make strong and timely tactical decisions.

As an intermediate emergency medical technician (defibrillator certified), I have the ability to adapt from the finger stuck in a pipe incident to the major medical incident. As a member of our haz mat team, I know when to slow down and smell the roses to gather critical information before tactical decisions are made. I am also a member of the high angle rescue team, confined space rescue, and wildland urban strike team. I have demonstrated my ability to take this educational knowledge and practical experience and apply it in practical terms for the betterment of this department. Some of my programs and projects include the rewrite of our

Fig. 4–5 Sample Response 3

Incident Command procedure, the development of our Field Operations Guide, the development and implementation of our Personnel Accountability Safety System, our Pager Recall system, stronger truck company operations, and co-development of our wildland urban interface strike team.

In the area of training, I have used my experience and education to develop a dispatcher training program, I am a CPR instructor–trainer, teach each year in our academy, and I place a very high priority on promotional training and have developed training programs for firefighters wishing to take promotional tests. As a captain, I will be able to further enhance our promotional training program.

In fire prevention, I am certified as a fire company officer fire code inspector through the International Fire Code Institute. I always take an active role in public educational opportunities. I have given many presentations on the fire service and its roll in the community before groups like Rotary and Lions Club.

Leadership is very important. When I was 18, I was the youngest assistant manager for a regional retail outlet. At 20, I was a forestry engine foreman. I have always embraced additional responsibility and leadership opportunities. As a captain, I will be charged with accomplishing our department's goals through my crew. I always set a positive example and over the years I have strengthened my leadership skills by improving my listening skills, acting in a consistent manner, expanding my knowledge base, and developing my communication skills. A captain must be open, honest, and fair, and I am.

In administration, my managerial and organizational skills have enabled me to develop and maintain the complex fire management reporting system that we use. I learned the system, developed and wrote the procedures and operational manuals, trained the department and developed statistical queries that lead to our current annual reporting system. I have also

Fig. 4–5 *Continued*

developed our records management program to meet state requirements for record retention, and record destruction. These skills will be applied to the management of my district responsibilities.

In each of these five areas of responsibilities for captain, I have demonstrated my desire, ability, and perseverance to get the job done in a goal oriented and positive manner. As your newest captain, I will have additional authority and opportunity to continue to give back to this department in a positive goal oriented manner and make a greater impact on the future of this department while meeting my responsibility to generate a better lifestyle for my family.

Fig. 4–5 *Continued*

Considering the concepts of describing the position for which you are interviewing, overviewing your qualifications, and relating your qualifications to the position we have discussed thus far, what did you think of this response to the opening question, "What are your qualifications?" To answer this question, let's analyze Response 3:

- Initially, let's briefly discuss a subject that is often a hidden factor in interviews and/or promotional advancements and is often referred to as, "What have you done for us lately?" When comparing this response to the previous response by the engineer candidate, did you notice the captain candidate has more fire department involvement and qualifications to discuss with the interview board members? This is normally due to more time in a department, and more involvement within the department (i.e., committees, projects, special assignments, etc.), and becomes more apparent as a person promotes upward through the ranks.

  However, this fact can also be stated from another perspective: "As your involvement within your department increases, so will your reputation to your department supervisors/chief officers. Therefore, your ability to discuss your specialized qualifications that are applicable to the position for which you are being interviewed also increase." Now, allow me to state that last sentence from another perspective: "With all things being equal, if you lack involvement within your department, don't expect to compete with the serious contenders who have been involved in your department and are also competing for the same promotion for which you are interviewing."

  An additional word of caution. If possible, try to keep a balance between off duty and on duty involvement. There are advantages with off duty involvement (i.e., community service), but be very careful about expounding on your off duty work for extra income (particularly when talking to a civilian board member).

- In the first and second paragraphs, the candidate briefly answered the opening question and then told the interview board what the position of captain was comprised of.

- In the first paragraph, this candidate boldly stated, *I am your best qualified candidate.* Some candidates are comfortable with this approach, but this author is not. If you tell the board you are the best candidate, then that is what they are expecting. However, if you are unable to back up your statement, then your final score might not meet your expectations. I clearly remember conducting an interview for the position of captain where a candidate confidently walked into the interview room, seated himself after being introduced to the three board members, and confidently stated, *Members of the board, during the next 40 minutes I will give you a 96 interview.*

My first thought was, *Oh really? Let's see.* During the next 40 minutes, this candidate was perfect, and I have to confess, it got to the point we could not ask a question the candidate didn't know. At the conclusion of the interview, the board chairperson asked the candidate if he had anything else to say.

The candidate again confidently, boldly stated, *Members of the board, during the past 40 minutes, I delivered a 96 interview, and I am your best qualified candidate for the position of captain.* After the candidate left the interview room, I turned to the chief next to me and said, "What did you think about that candidate?" The chief answered, "Awesome. He was a 100." I then said, "So what are you going to give him?" He answered, "A 96, that's what he asked for."

- In the first paragraph, the statement *so I may utilize my education and experience to make a greater impact on this department* is excellent, and is also a very true statement. However, the statement would be more inclusive if it was worded *so I may utilize my qualifications, experience, and education to make a greater impact on this department.* As you promote through the ranks, your department is giving you the opportunity to use your qualifications to impact a greater span of control (or more people) for your department.

- How many times in the response did the candidate use the term *as a captain*? The answer is *not enough* (if you count, it's three). However, in the last paragraph, the phrase *as your newest captain* is used. This is excellent as the candidate is speaking *from* the position of captain.

- How were the qualifications related to the position of captain? To answer this question, let's take a closer look. Was the position of captain directly related to qualifications under the headings of operations and fire prevention? The answer is no. However, qualifications were somewhat related to the position of captain under the headings of training, leadership, and administration.

  In the final paragraph, the candidate used the phrase *in each of these five areas of responsibilities for captain, I have demonstrated my desire, ability, and perseverance to get the job done in a goal oriented and positive manner.* Although this approach is good, this author prefers to relate qualifications under each area of the position for which you are interviewing. When you deliver your opening response, remember to relate, not somewhat relate.

- Notice that the last sentence states the candidate needs to *generate a better lifestyle for my family.* I realize I am on thin ice with this one, but consider what the last thought of this response stated—money! Although there is nothing wrong with this concept, be careful how you address this issue (if you address it) and don't make it a primary consideration.

- Last but not least, this candidate has numerous qualifications that are excellent but were somewhat applied to the position of captain.

In this response, the area of operations definitely has the most qualifications, with leadership next, then training. The area of fire prevention has the least amount of qualifications.

The point is, the area of operations has a tendency to overshadow the other areas. If possible, try to balance your qualifications under each area of responsibility. For example, under the area of operations, the last section *some of my programs and projects include...* could have been moved to the area of training that would have decreased the length of operations, and increased the length of the area of training.

Battalion chief

Finally, let's look at the position of battalion chief and how we can divide it into its multi-dimensional elements. Although the horizontal elements are essentially the same as sergeant, lieutenant, and captain, the primary difference of management and responsibility will be noted in the dimension of vertical elements. The position of battalion chief can be divided into five job elements (horizontal dimension):

1. **Department experience/operations.** Consists of actions necessary to mitigate all types of emergency and nonemergency incidents.

2. **Fire prevention.** Consists of actions necessary to ensure that inspections, activities and enforcement of applicable laws and codes are completed in an acceptable and timely manner.

3. **Leadership.** Consists of the ability to direct assigned resources under your command/supervision. It can also include your accomplishments as an officer.

4. **Training.** Consists of a combination of ensuring appropriate training programs are conducted, developed, and implemented for and by your command, and can also include your educational achievements.

5. **Administrative.** Can be utilized as a "catch all," and can include items such as records, reports, goals, routine operations, and most importantly, public relations/customer service.

Again, notice that these five areas encompass the main areas of responsibility for the position of battalion chief and will vary depending on the specific job description and your opinion. However, they are your opinion, they begin to form an outline and a starting point for this position, are easy to remember in an interview environment, and are similar to the headings on the resumés in chapter 3. Next, list all of your qualifications under each area similar to the following examples (vertical dimension):

- **Operations**

  ✓ 20 years of experience

  ✓ obtained ranks of engineer, lieutenant, and captain

  ✓ incident commander

  ✓ acting battalion commander

- **Education**

  ✓ Associate of Arts in Fire Science

  ✓ electronic engineering

  ✓ public administration

- **Leadership**

  ✓ event chairperson firefighter olympics

- **Training**

  ✓ department training programs

  ✓ author fire service magazines

- **Administrative**

  ✓ administrative office

  ✓ department spokesperson

  ✓ member emergency operations command

As this outline applies to the following sample response, notice this candidate used operations instead of department experience, fire prevention was deleted, and education was substituted for fire prevention. As a result, operations was used with education, leadership, training, and administrative. Also notice in the answer that this candidate chose to discuss operations, administration, training, education, and leadership in that order. The order and selection of these specific areas are not important. What *is* important is that each candidate selects the areas that apply to appropriate individual qualifications, as qualifications will vary from candidate to candidate. The order you discuss the areas of qualifications in the interview is also not as important as the order you are most comfortable with.

Now, consider the common opening interview question, "What is the position of battalion chief?" Response 4 utilizes the multi-dimensional overview (Fig. 4–6).

Sample Response for Battalion Chief

Members of the board, the opportunity to work for this department has given me numerous opportunities to serve this department and the citizens of the community and also evaluate and apply the various assignments and opportunities made available to me. Let me take a few minutes to overview the position of battalion chief and my qualifications for this position.

The position of battalion chief has five major areas of responsibility: operations, training, education, leadership, and administration.

In the area of operations, I have 20 years of varied experience in field operations and administrative assignments. I have promoted through the ranks and have worked approximately five years in each position of firefighter, engineer, lieutenant, and captain. A basic understanding and knowledge of each rank will assist me as a chief officer in knowing the capabilities and potential of each position. My assignments have been in areas classified as high-risk, high value, high incident, commercial, and residential. I have utilized these assignments to establish and expand my knowledge of engine and truck operations, and single and multi-company operations. As an incident commander, I have utilized the effective deployment of these resources to mitigate various types of incidents. This will assist me as a battalion chief in the utilization and deployment of these types of resources to handle incidents in concert with the incident management system.

The opportunity to act as a battalion chief has enabled me to gain practical experience as a battalion chief/incident commander at various types of incidents. As a battalion commander, I have successfully commanded large incidents that utilized my ability to evaluate specific needs to operate within the incident management system, deploy resources, communicate with resources, and other required functions necessary or mandated by emergency incidents. Acting battalion commander has also enabled me to conduct periodic inspections of a battalion that have

Fig. 4–6 Sample Response 4

developed my ability to evaluate a command, assume responsibility for administrative functions such as records and reports, plan and implement battalion training sessions, assess the overall efficiency of a battalion, and direct the coordination and efforts of the people and resources within a battalion.

In the area of administration, my current assignment to administrative duty has enabled me to work within the department's administration and as an integral part of the chief's staff. This has given me the opportunity to strengthen my administrative skills that are fundamental to the position of battalion chief. Working within the administration has also familiarized me with each section and its personnel, and will facilitate knowing where to go and who to ask for information that is necessary to administer a battalion and/or keep it properly informed. I have represented the department before other city departments such as recreation and parks, personnel department, and building and safety. This has sharpened my communicative skills that I will use in presenting this department and establishing channels between my battalion and local civic groups necessary for required community and public relation activities. My participation in the city Emergency Operation Center during the city's recent disaster has given me the opportunity to coordinate emergency operations between this department and other city departments.

My assistance in developing mutual and automatic aid agreements with other fire departments has developed my ability to evaluate, implement, and follow-up on necessary activities. I will utilize these skills to evaluate, organize, implement, and monitor battalion programs. Developing videotape newsletters and material for national presentations by the chief has demonstrated my ability to plan and develop practical and effective information that is utilized on a department and national level. This will assist me in the development–effective presentation of relevant material within my battalion and also be able to sell and implement departmental programs.

Fig. 4–6 Continued

My experience in the planning and research section has enabled me to develop and recognize administrative reports and correspondence that utilize correct formats and incorporate accurate and concise information. This will enable me to utilize and set the standard for reports and records within my battalion. My ability to research, evaluate, and submit recommendations for indepth studies to assist the chief in planning and achieving goals will enable me to prepare accurate and concise staff reports for my supervisors.

Working within the administrative office has acquainted me with the administrative process as applied to the operation and future direction of this department. I will utilize this information in directing the present and future course of my battalion. This is particularly helpful when determining trends, problems, and factors that affect the battalion/department and my evaluation and recommendations to my people and supervisors.

In the area of training, I have developed and given training programs to this department that have resulted in my developing and producing department video training tapes and material. This expertise has enabled me to develop and author fireground articles for national publications that have led to national recognition of effective training material. This has given me the opportunity to give training programs to other fire departments as well as participating as a speaker at national fire conferences. My involvement in these programs has demonstrated my ability to focus and identify a need, and then initiate, plan, and develop–implement successful training programs. This has developed my ability as an effective credible communicator and expanded my ability in working with and dealing with people. As a battalion chief, this demonstrated performance will assist me in two major areas. First, I will be able to plan, organize, direct, and evaluate effective training programs within my battalion and with this department. Secondly, my ability to communicate with people will be utilized in representing this department and integrating my battalion with various homeowner and civic groups (effective and practical community and public relations).

Fig. 4–6 Continued

Education does not guarantee a successful chief officer; it will build a stronger foundation and increase technical expertise. I have received an Associate of Arts degree in fire science and have substantial units in electronic engineering and public administration. This background will give me the extra tools and resources that I will utilize to run an effective battalion.

Finally, leadership is the cornerstone of the qualifications for a chief officer. My recent coordination of an event within the Firefighter Olympics that involved more than 300 firefighters from this state underscores my initiative to see a need, initiate a plan, and then provide the necessary leadership to ensure its successful completion. The success of this program depended on my ability to evaluate the needs of the program, plan (short and long range) the event, organize necessary resources, direct and control those resources, provide clear and effective communication, ensure correct reports, and operate within an allotted time frame and budget.

Members of the board, each of these areas are the same tools that I will use to run an effective battalion as a battalion chief in this department.

Fig. 4–6 *Continued*

Again, considering the concepts of (1) describing the position for which you are interviewing, (2) overviewing your qualifications, and (3) relating your qualifications to the position that we have discussed thus far, what did you think about this response to the opening question, "What is the position of battalion chief?" To answer this question, let's analyze Response 4:

- In the first and second paragraphs, the candidate briefly answered the opening question and then summarized the position of battalion chief. When this candidate has finished with the interview, will the interview board members be able to confirm this candidate defined the position of battalion chief and applicable qualifications? Absolutely!

- How many times did this candidate use the terms *my battalion, battalion, as a battalion chief*, etc. The answer is about 20 times. This candidate consistently tied qualifications to the position of battalion chief throughout the response. However, notice the candidate did not relate any qualifications to the position of battalion chief under leadership. Oops!

- During this response, did this candidate *consistently* speak from the position of a battalion chief? Absolutely!

- Similar to the engineer and captain responses, each area of responsibility is initiated with the name of the specific area of responsibility (i.e., *in the area of training, education, and finally leadership*, etc.). This approach clearly delineates each area of responsibility, is easy to remember when overviewing your opinion of the position, and your applicable qualifications.

- Currently, the higher you promote in the modern fire service, the more important education becomes. In this response, the candidate has an Associate of Arts in Fire Science and units in electronic engineering and public administration, which is good but not the same as a Bachelors in Public Administration. A method to enhance your current education level is to talk about future education goals. For example, *my current education consists of an Associate of Arts degree in Fire Science; however, my goal is to obtain a Bachelors of Arts degree in Public Administration within the next five years.* Obviously, you would only use this approach if it were a true statement.

- Briefly consider the comments in the area of leadership. Notice the candidate was involved in coordinating an event in the firefighter olympics, but did not specify the particular event. The specific event is not as important as what the candidate can apply from coordinating the event. The same concept applies to candidates who can talk about responsibilities in a church or other volunteer environment. The type of organization is not important, but what you can apply from your specific responsibilities is important.

Additionally, there is a major difference between leadership in the fire service, and leadership in a volunteer environment. When a company officer gives an order, it must be satisfactorily accomplished or there will be disciplinary consequences. However, when an order or suggestion is given in a volunteer organization (church, Rotary club, etc.), the person receiving the order can either obey or leave. The moral to this story—if you can display effective leadership in a volunteer organization, you can lead in the fire service.

- The phrase *demonstrated performance* can be used for a specific emphasis and deserves your consideration. As we have previously discussed, if you talk about a qualification, what gives that qualification credibility? If you can give examples of what you have done in the area of your qualification, then your qualification becomes credible. If you tie the phrase *demonstrated performance* to your qualifications and examples, you can enhance credibility and focus the board members' attention to this fact. For example, see the use of demonstrated performance under the section on training in this response.

- Finally, this answer would take about 3½ to 4 minutes to deliver in an interview setting. Try not to go more than 4 minutes with your answer to an opening question. Although time flies when you are having fun, interview boards are normally limited to a tight time schedule.

The preceding concepts we have discussed in this chapter to this point will assist you in two additional areas:

1. If it is difficult writing out an opening statement for your interview, the outline format we have discussed will simplify that procedure.

2. If you are not able to give an overview of your qualifications or overview the position at the beginning of the interview, any questions that relate to these areas will be easily answered.

Common Opening Questions

Let's consider some examples of common questions often used during the opening portion of an interview:

- Can you give an overview of yourself?

- Can you summarize your background?

- Where have you worked?

- What are your qualifications?

- Why do you want to be a _____?

- What are you doing now that qualifies you for _____?

- What makes you better qualified for the position of _____ than the other candidates?

- How does your experience and education qualify you for the position of _____?

- What can you tell about yourself?

- What strengths do you have that qualify you for the job of _____?

- In your opinion, what is the job of _____?

These questions "give you the ball," and let you run in any direction you choose. If you simply answer these questions and let the interview board infer your background has given you the necessary qualifications, you are in trouble. However, if you can describe the position you are seeking, overview your qualifications and relate what you have learned to the position you are seeking, you will have maximized your answers to their question(s).

Although these questions are similar, two are unique. The question, "What makes you better qualified than the other candidates?" is answered by stating two facts. First, you are not familiar with the qualifications of the other candidates, and second, you *are* familiar with your qualifications. The answer to this question is *I am not familiar with the qualifications of the other candidates. However, I am familiar with my qualifications. Let me take a few minutes and overview my qualifications and the position of _____.*

The question, "Why do you want to be a _____?" will give the interview board members an opportunity to analyze your thought process and listen to your opinion of why you will be an effective officer. Additionally, it will provide the opportunity to evaluate your *honesty*. Do you want the new position because you are uniquely talented and a wonderful person, or because you are a qualified and effective candidate to offer to your department? Start by asking yourself why you really want the position, as you are the only person who can answer that question. Some valid reasons are as follows:

- You have the qualifications and demonstrated performance for the new position (this is what your department is looking for).

- Your department needs strong effective leadership (you are that person). Your demonstrated performance has echoed that fact.

- You will have an opportunity to impact and affect more people for your department. Generally, each promotion you make will give you the opportunity to impact more people. If you have the qualifications for which your department is looking, they will give you that opportunity.

- A promotion will raise the standard of living for your family. Now we get to the honesty part of this question. It is no secret that promotions equal more money. This translates into a higher standard of living for you and your family. This consideration is a fact of life, practical, and you are being honest. However, if you use this perspective, don't make it a priority. An answer to this question could parallel the following:

> *I desire to utilize my qualifications in the position of captain, engineer, battalion chief, etc, to impact and affect more people for this department with my qualifications in connection with providing strong and effective leadership. My demonstrated performance in these areas has underscored my ability to accomplish these goals. Additionally, I have the opportunity to raise the standard of living for my family.*

Remember, these questions set the standard for the interview—be prepared!

Tips for Delivering a Response to the Opening Question

- Be comfortable. Sit naturally and use your hands if that is your style.

- Limit your opening response to not more than four minutes.

- Maintain eye contact with each member of the board during every question.

- Speak at a normal pace.

- Smile and be confident.

- Put yourself in the position you are seeking.

- Memorize general ideas, not your exact opening.

- Wipe your introduction hand on the appropriate part of your clothing just before you enter the interview room. If your hand is wet, this will remove any moisture and stimulate circulation.

- Stress positives, never negatives.

- Get adequate sleep the night before the interview (if possible). This also means not working the day before your interview as you never know when you will be up all night with a major emergency. Take the day off as you have too much invested not to be your best. Additionally, monitor your food and liquids the day before the interview.

Summary Checklist

1. Generally, an interview board will put a candidate at ease by starting with an opening question that will allow the candidate to talk about strengths, qualifications, background history, and other similar subjects.

2. During opening questions, candidates should consider answering the question, overview applicable qualifications, overview the job, and relate qualifications to the job.

3. Any job can be considered multi-dimensional:

   • Job elements (horizontal dimension)

   • Candidate's qualifications (vertical dimension)

4. To develop an initial outline, candidates should separate a position into four or five elements, and list appropriate qualifications under each element.

5. Opening statements should not exceed approximately four minutes.

5. General Questions

Introduction

As we discussed in chapter 2, a promotional interview is comprised of three main parts:

1. Opening

2. Body

3. Closing

Once the interview board and candidate have completed the opening portion of the interview, it is time for the fun portion of the interview—the body. Although the opening portion normally utilizes one or two typical questions that can be designed to give the candidate an opportunity to display appropriate qualifications and an understanding of the position, the body portion will utilize a wide variety of types and styles of questions that are designed to

- gauge your ability to think on your feet

- analyze your thought process

- display your opinion on various subjects

- demonstrate your knowledge of your department's practices, procedures, policies, and other similar subjects

However, before attempting to answer questions during a promotional interview, remember the following basic considerations:

- Don't ever tell an interview board what you think they *want* to hear. Either you have the confidence to give your opinion and/or make a decision on any subject, or you do not. Remember, an interview board is evaluating your thought process.

- Keep answers as short as possible while answering the question. An interview board is extremely appreciative of a candidate who can answer a question once while keeping the answer short. Most candidates will answer a question, answer it again from a different perspective, then answer it again just to ensure the board didn't miss the answer! When you videotape your practice interviews, evaluate this important criteria in your answers.

- Don't use abbreviations/acronyms or fire department lingo (i.e., task force, IMS system, EMS, etc.) if a civilian is on your interview board, and be careful if members from other fire departments are present.

- If the interview board asks you the same question from a different perspective after your initial answer, you probably did not answer the question.

- The duration of your interview is not an accurate gauge of your performance/final grade. For example, let's assume your interview is only 20 minutes long. It is possible for the board to conclude your interview for two reasons:

  1. You have done very well and answered all of the necessary questions, so why continue the interview?

  2. It is obvious you need more experience and have not done well with the necessary questions. Again, why continue the interview?

You will know the answer to this simple question when your final grade is posted.

- Don't ever walk into an interview and hope the board doesn't ask you a question in a specific area. Murphy's Law says, "If that is the case, they probably will ask you the question you are dreading." Ensure with proper prior study you have eliminated this concern before your interview.

- Occasionally, a board will ask a candidate a "teeter-totter" question, which is normally unanswerable. This is designed to test the candidate and/or see what the candidate will do. If you are asked a question to which you don't know the answer, it is no big deal to just say, *I don't know the answer, but when I leave the interview, I will find the answer.* However, if you have to say I don't know three or four times in an interview, you are in trouble.

- This recommendation is not rocket science, but make sure you absolutely and positively know all of your department's requirements that you would be responsible for in your new position.

- Don't flip flop on an answer unless you absolutely know you made a mistake. There are times when a board member might challenge an answer to see if you are confident in your answer. If you change your mind because you think the board member wants you to, or you feel your answer *might* have been wrong, you are toast! If you are confident in your abilities, don't waffle.

- Don't talk about your department in negative terms. Be positive (loyalty), but be willing to discuss any appropriate areas that you feel can be improved (and be sure to have a solution).

Questions

Due to the wide range of various questions that can be utilized, let's consider some typical questions for which you can anticipate and prepare. Let's discuss the position of engineer for a moment. The position of engineer is significantly different from supervisory positions because of the technical nature and requirements of the position. It should be anticipated that a candidate for engineer would be questioned on pump theory, aerial device theory, operational requirements for pump and aerial devices, apparatus placement, and so on. Therefore, it is imperative that a candidate for engineer is prepared to discuss any area of appropriate responsibility, and specifically the following considerations:

- Pump theory

- Fire service hydraulics

- Aerial device (ladder/platform) theory

- Aerial device operational considerations

- Building construction

- Apparatus placement considerations at incidents, specifically, the relationship of trucks and engines (if this applies to your responsibilities).

- Fireground operations (engineers are responsible for having a working knowledge of fire attack, ventilation, search and rescue, forcible entry, 2 in/2 out, rapid intervention team [RIT], etc.).

- Engineer's role in community service. For example, when an engineer drives fire apparatus, either from an emergency or nonemergency perspective, that engineer and apparatus represents that fire department to the community it serves. How do you represent your fire department?

- The opportunity an engineer has to train firefighters (under the officer's guidance) in the proper driving, operation, and maintenance of fire apparatus.

- Interpersonal relationship. For example, a question that would parallel, "You are appointed to engineer on an engine company, and after several months you notice that one of your reliefs is not properly maintaining the apparatus. What would you do?" The answer is, *I would notify my commanding officer of the facts and let my officer handle the problem with the commanding officer that is responsible for my relief on the other shift.*

- Lastly, consider being able to discuss with the interview board any ideas that you have regarding changes/improvements/expertise you could make/recommend as an engineer. For example, due to your background and interest in driving and maintaining heavy apparatus, you would like to design and operate a driving and maintenance academy for current and future engineers on your department.

Now, let's discuss some general questions that should be anticipated in a promotional interview. The following questions will not be listed in any particular order, can be applied to numerous types of positions with some thought and appropriate modification, and will cover the main areas of questioning.

What are your strong points?

This question is designed to let candidates express opinions of themselves. It also gives an opportunity to sell the qualifications that you think are important to the position (and the same qualifications you possess). Additionally, although this sounds like an easy question (and it can be), most candidates are not prepared to discuss several of their strong points and then relate (there is that word relate again) them to the position they are seeking. Consider the following examples:

- Desire. Your desire is to make your command the best it can be.

- Decisiveness. You are a decisive person.

- Initiative. You demonstrate initiative in starting and accomplishing goals.

- Effectiveness. You will run an effective command.

- Attitude. You have a positive attitude toward your department, your promotion, and what you will be able to accomplish.

- Communicate. You are able to effectively communicate with other people (don't use this example if your interview is not going well).

- Planning and training. You will plan future goals and ensure a high level of training for your command.

- Leadership. You get the job done through people.

- Sell and implement your department's programs. You can be depended upon to support and implement necessary programs.

Obviously, you should choose strong points that apply to you. Additionally, be prepared to expand on any of your attributes, and *relate* them to the position for which you are interviewing. For example, assume you choose to discuss planning as one of your strong points. You could say,

The ability to plan and set appropriate priorities is one of my strengths, and is critical to the qualifications of an effective officer. As the board is

aware, an officer must be able to set the priorities for each shift and be able to adjust them as necessary. As an officer, I will place a strong emphasis on planning and setting the priorities for each shift and my command.

Now that we have considered the concept of strong points, you should also be able to discuss some of your weak points (and everyone has them). Be honest and analyze your weak points; have several that you can discuss. The focus of this concept is to choose a weak point, discuss the weak point, and then turn it into a positive. For example:

I believe that my two-year degree in fire science will not be sufficient for future promotions. I plan on obtaining a degree in business administration to enhance my business and administration skills.

Be sure you don't discuss a weak point without a positive cure!

What is your definition of leadership?

Although this area may or may not be discussed, it is definitely worthy of your attention due to the fact that leadership is an essential quality in any officer, and to varying degrees, positions below the rank of officer. Consider the following questions:

- What is your definition of leadership? Webster's dictionary defines leadership as *to guide, to direct the course of by going before or alongside,* etc. The point is, what is your definition?

- What are your qualifications and demonstrated performance in the area of leadership? This question can be easily answered from your resumé/opening question preparation.

The following 10 principles of an effective leader are presented for your consideration:

1. Trusted–trustworthy.

✓ Trusted by others and friends. Do people believe in you?

✓ Do they have the confidence that you will have their best interest at heart?

✓ Do you lead from the vantage point of service believed in?

2. Leaders always take the initiative.

✓ Identify the need.

✓ Have a solution.

✓ Take action.

✓ Delegate responsibility and work alongside until it's right.

3. Utilize good judgment.

✓ Cautious.

✓ Careful.

✓ Sensitive.

✓ Seek wise counsel to derive wisdom.

4. **Speak with authority.**

✓ Boldness. Know your subject and mean it.

✓ Confidence. Predicated on the knowledge of truth.

5. **Strengthen others.**

✓ Build on strength. Improve weaknesses.

✓ Upward mobility (training for promotion).

6. **Optimistic and enthusiastic.**

✓ Positive outlook.

✓ Excited to be involved.

7. **Never compromise absolutes.**

✓ If you compromise your standards, you lose your integrity and your following.

8. **Focus on objectives, not obstacles.**

✓ You move around obstacles to get to the objectives.

9. **Lead by example.**

✓ Get out front and set the pace.

10. Activate people.

✓ You involve the people under your command and delegate authority and responsibility.

Remember, a supervisor manages things and leads people. Additionally, 97% of a line officer's job is managing resources (most of which are people) in a nonemergency environment. The other 3% is emergency incidents.

What is your definition of loyalty?

If an interview board member wants to ask a question and receive a blank stare from a candidate, this is the question. To answer this question takes prior thought before walking into an interview room. One suggestion is to divide loyalty into three areas:

1. Your department

2. Your supervisor

3. Your subordinates

Your answer could then be, *As a captain, loyalty is my support of this department, my supervisors, and the people in my command.*

General knowledge

A general knowledge of current events (within the following areas) can give you the opportunity to overview these events, and demonstrate your current knowledge of factors affecting the fire service. These areas are summarized as

- national

- state

- local

- your department

Although it is not necessary to be an expert in these areas, be aware of organizations and events that affect and shape the fire service, and consequently, your job. You should have a basic understanding of the following areas:

- **National**

  ✓ Federal Emergency Management Agency and the National Fire Academy.

  ✓ Current events and legislation. For example, in the current events section of a popular fire service magazine, three pending legislation issues are detailed:

    1. Assistance to Firefighters Grant Program (FIRE Act)

    2. 700 MHz for Public Safety

    3. Public Safety Officer's Benefits (PSOB)

- **State**

  ✓ Your state fire marshal and state fire academy.

  ✓ Current events and legislation.

- **Local**

  ✓ Any current city policies and mandates affecting your department.

- **Your department**

  ✓ Your department budget.

  ✓ New policies, programs, directives, etc.

  ✓ What direction is your department going?

  ✓ What are the goals and objectives of your department?

  ✓ Are you familiar with the *Mission Statement* of your department?

  ✓ What are current department problems, concerns, and what ideas and solutions do *you* have?

What are your goals and objectives?

A quality demonstrating your ability to plan and influence your future (and your impact to your department) is the ability and desire to evaluate and improve your performance and job position. You cannot gauge your effectiveness unless you can measure your progress (Fig. 5–1). This is accomplished by defining, setting, implementing, monitoring, and evaluating goals and objectives. An excellent question to a perspective candidate is, "When you step into that new position, what are your goals and objectives?"

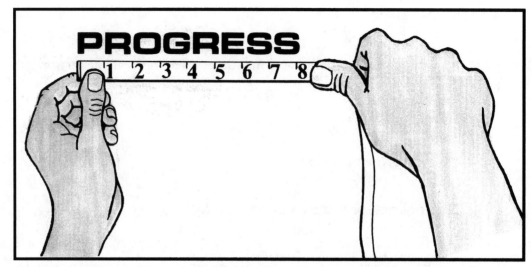

Fig. 5–1 You Cannot Gauge Your Effectiveness Unless You Can Measure Your Progress

Goals and objectives can be divided into two categories:

1. Personal

✓ What are your personal goals and objectives?

2. Command

✓ What are your goals and objectives for your command?

Additionally, consider separating goals and objectives as follows:

• **Immediate**

✓ Where are you now?

✓ What do you have to work with?

✓ What is your/their current level of performance?

- **Short range**

  ✓ Four to six months. What do you want to accomplish during this time frame?

- **Long range**

  ✓ One to five years. What do you want to accomplish during this time frame?

These time frames will vary according to your opinion. A realistic approach in this area will demonstrate your maturity and desire to evaluate the present performance and potential of yourself, your command, and your initiative to improve and gauge your effectiveness in accomplishing these considerations. Implementation of goals and objectives will enable you to determine if your expectations have failed or succeeded, and if additional fine tuning and changes are necessary.

Emergency incidents

Questions that fall under this area are answered from a foundation that is governed by three parameters:

1. **Standard operating procedures, and policies–procedures.** When considering problems such as structure fires, haz mat, brush fires, emergency medical services (EMS), and other similar incidents, some departments have specific operational procedures for specific incidents, while other departments have general or no procedures. Obviously, if your department has specific operational procedures for specific incidents, this will affect the way you answer an applicable question. You should be totally familiar with your operational procedures.

2. **General or specific.** Questions will either be specific or general. Determine if the interview board is looking for a broad approach concept (overview) or for specific information. Let's consider two examples:

• "Describe your actions as the first officer to arrive at the scene of 30-story high-rise building with fire reported on the twentieth floor" (Fig. 5–2). This question is looking for a general course of action to abate an *apparent* problem.

Fig. 5–2 Describe Your Actions at a 30-Story High-rise Fire

- "Describe how you would extinguish a fire involving blocks of magnesium." This question is looking for a specific course of action to abate a specific problem.

Utilizing the preceding information as a foundation, consider the following when formulating your answer:

- When applicable, utilize your department operational procedures. In this case, you are being evaluated on your knowledge of your department's applicable procedures.

- Determine whether the question is general or specific.

    ✓ If a question is specific, you must provide the requested information.

    ✓ If the question is general, you have more leeway in your answer. However, most questions are simplified by the use of concepts, not details. Details tend to prolong your answer, lead to tangents, and invite the board to key on your specifics. Remember, if a board wants to elaborate on what you have said, they will!

3. **Set the rules before you answer the question.** Consider the following question: "Describe your actions as the first officer to arrive at the scene of a four-story apartment building with fire showing from the third floor."

Notice the wide latitude in this question and the information that is missing:

- As the first arriving officer, what type of company are you assigned to?

- Is fire visible from one window or the hallway?

- Is smoke visible from the fourth floor?

- Are there people in the windows?

- Is it noon or midnight?

These considerations can dramatically alter this question. Before you answer this question (or similar questions with questionable information), it is imperative that you and the board perceive the same problem. Because people are human, it is possible for a board member to state a problem differently than it is perceived in the mind, or the board member may have just forgotten several items in the question. However, it is possible to modify or simplify a question and ensure you and the board perceive the same problem. Consider the following response to the previous question:

I perceive this problem as initially placing my company on the scene of a four-story apartment building with fire showing on the third floor. Additionally, let's assume that one unit is well involved on the third floor; smoke is showing from the fourth floor, it is 1000 hours, and I am the captain of an engine company.

Unless the board interjects additional information, you are ready to answer the question. It should be apparent that, if appropriate and within certain limitations, you can:

- simplify or modify a question.

- alter a question to suit your answer.

- make it difficult for the board to interject additional information during your answer.

- eliminate possible confusion between you and the board in perceiving the same problem.

- give yourself a few more seconds to analyze the problem before giving an answer.

- demonstrate your confidence in answering the question.

Briefly, let's consider the concept of answering a question with specifics or general concepts. Utilizing the preceding question, consider the following responses:

• Specifics.

I would lay a 4-inch supply line from the nearest hydrant, give a size-up, direct my personnel to utilize full protective equipment, put a ladder to the second floor fire escape, and take a 1¾-inch line to the third floor to extinguish the fire, etc.

Although this answer is correct, the response has utilized specific information that can be replaced by information an incident commander (you) should be concerned with (remember the board made you the first-in officer. Take yourself out of the locker room and put yourself in the position for which you are interviewing.).

Now, let's consider this question from the perspective of general concepts:

• General concepts.

As the incident commander, I would provide an initial size-up and request additional resources as necessary. I would direct my company to take a line to the third floor and initiate an attack on the fire. I would also utilize additional companies to check for extension on the fourth floor, affect ventilation if applicable, back up my company on the third floor, ensure search and rescue operations are completed, check for possible salvage operations on the second floor, etc.

With this approach, you have utilized concepts to mitigate this incident from the perspective of an incident commander. The method used to answer a question must fit each type of question and your style of applicable tactics.

When answering these types of questions, an easy way to ensure that all sides of an incident are covered is to use the "six-side approach." For example, assume a fire on the second floor of a three-story building (Fig. 5–3). Using the six sides:

- Fire attack and backup lines on the second floor. This covers the perimeter of the fire floor, or the four sides of the fire floor.

- Extension and ventilation on the floor above (fifth side).

- Salvage on the first floor (sixth side).

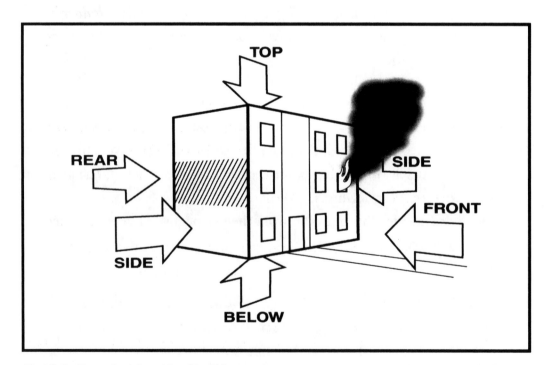

Fig. 5–3 Every Incident Has Six Sides

As an additional example, assume a gasoline tanker has collided and exploded against a bridge abutment:

- Confinement and exposure protection for the perimeter (four sides).

- Gasoline flowing into storm drains (fifth side).

- Where is the smoke cloud going (sixth side)?

Briefly, let's consider the IMS. If your department utilizes an IMS system, are you totally familiar with its operation, and are you able to apply and expand it to any type of incident? Additionally, can you define the IMS system? Consider the following:

- The IMS is a Standard Operational Procedure (SOP), or a prefire plan of an organizational structure to mitigate an incident. It is adaptable to any type of incident, and is expandable to meet changing conditions and specific needs. Resources should know what needs to be done when inserted into the system as each position has pre-planned responsibilities.

- Be familiar with the concept that under the IMS system, the incident commander (which initially includes the first-in officer) retains all responsibilities unless they are delegated, and the IMS system is *always* operational. It is not a system that is turned off and on.

- The flexible nature of the IMS system allows an incident commander to activate only those portions of the IMS system necessary to mitigate an incident. As the incident grows, so can the IMS.

Within the area of the IMS, an excellent question to ask a candidate is, "Are you familiar with the IMS?" Naturally, the answer is always a resounding *yes*. The following question from the board will then be, "What is the/your definition of the IMS?" or, "Can you describe/explain how you would use the IMS?" If you cannot answer these questions, the previous *yes* answer has little credibility.

Ensure that you have given thought to the mitigation of any type of incident you could respond to. Some examples are as follows:

- structure fires

- multi-casualty incidents

- brush incidents

- haz mat incidents

- EMS incidents

- general and heavy rescue incidents

- terrorist incidents

- aircraft/railroad incidents

- vehicle extrications

- floods/tornados

To enhance your familiarity and ability to answer questions for these incidents, it is recommended that you make an outline of your actions for each applicable incident. This will give you an outline for each incident, keep them fresh in your mind, and allow you to follow your outline in the interview.

How will you run a public and community relations program?

Without a doubt, both the immediate and long-range fire protection and medical service needs of your community are greatly dependent upon the understanding and support of the general public. Fire protection, fire prevention, and emergency medical service goals and objectives can be enhanced by each member of your department if they recognize and exercise their responsibility to properly condition the citizens they serve. Additionally, fire departments do not sell a product; they are paid to perform a service. However, we can sell our image. It is vitally important for you, particularly as a future officer, to develop an outline of goals and actions necessary to implement, maintain, or improve a viable public and community relations program for your command and department. Before we consider a few ideas, let's define our subject:

- **Public relations**. Public relations is public conditioning. This is the business of developing, changing, and enhancing a favorable public response in cooperation with the programs and needs of your fire department.

- **Community relations.** Community relations are the total of the attitudes, impressions, and opinions of your community in its relationship with your fire department. Hopefully, they do not mirror the person in Figure 5–4!

Fig. 5–4 Don't Give the Wrong Impression

Now, how are you going to accomplish the proper conditioning of *your public?* Let's look at a few considerations:

- **Attitude.** Develop in your subordinates an understanding and attitude of courtesy and objectivity toward the citizens with whom they/you come in daily contact. The individual member reflects the policy of your department. They are representatives of your entire department. This is accomplished by

  ✓ your setting the example

  ✓ training

  ✓ participation

  Positive contact with the public during fire prevention activities, fire prevention week, open house, the way personnel drive department apparatus (particularly in nonemergency operations), etc.

- **Contacts.** Develop contacts in your area with the following examples:

  ✓ Local newspapers

  ✓ Homeowners groups

  ✓ Civic organizations

  ✓ Schools

- **Media.** Utilize the media at emergencies in a positive manner, and don't forget their potential interest in noteworthy training programs.

What kind of a command are you going to run?

Before you begin to develop and maintain an effective command, you must first determine what you have to work with (Fig. 5–5). You should assess your new command by determining the level and quality of:

- **People**
  - ✓ Morale
  - ✓ Motivation
  - ✓ Standards
  - ✓ Performance

- **Resources**
  - ✓ Quarters
  - ✓ Apparatus
  - ✓ Equipment and supplies

- **Area**
  - ✓ Type
  - ✓ Needs
  - ✓ Specific hazards

Fig. 5–5 What Kind of a Command Will You Run?

With this basic information, you can determine any problems, needs, and priorities to implement applicable measures to achieve necessary standards. Remember that goals should be utilized to define your priorities and gauge your accomplishments. Also, consider how you will impact and institute any budget reductions that might be possible within your command. The following are possible areas:

- Conservation of supplies.

- Conservation of utilities.

- Injuries on duty. If you can isolate any injuries that are predominant or can be reduced through proper training, you may be able to save your department money.

• Apparatus.

• Applicable training or driver training programs may reduce accidents and improve apparatus longevity.

• Equipment. Can this area be improved or utilized more efficiently?

In the preceding paragraph, did you notice the word *standards?* This is potentially one of the most dangerous words you can use in a promotional interview (Fig. 5–6). Although this word sounds impressive, it is difficult to define unless you have given this word serious thought. Consider that anytime you use the word standards, you must be able to define what your standards are.

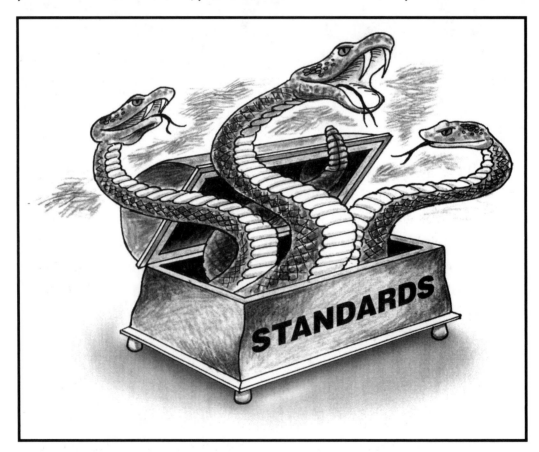

Fig. 5–6 Be Careful with the Word *Standards*

To illustrate this point, consider a question that is often used to determine if you have a practical grasp and understanding of the position for which you are interviewing. Let's assume you are interviewing for the position of lieutenant, and the question is, "What are you going to do the first day of your new assignment?"

A common response to this question is:

I will start my first day by reporting to my command early, checking my safety equipment, reading the journal of the previous day's activities, and meeting with my subordinates to discuss my standards. Then I will tour with my command to overview my district, familiarize myself with any applicable hazards, and conduct an afternoon drill to see what my command is capable of.

What did you think of this response? Was the response wrong? No. If your opinion was the response was shallow and lacked a realistic approach, you were correct. As I have watched numerous candidates use an identical or similar approach, I can guarantee the next question by a board member will be similar to one of the following examples:

- "What are your standards?" At this point, the candidate will be entering the early stages of shock!

- "Suppose your shift was previously scheduled for a drill in the morning at 8:30, and a school demonstration at 1:30 in the afternoon?" Now, the candidate has realized there is a major problem, and the interview is not going as planned.

Let's consider another response by a candidate for the position of lieutenant to the same question, "What are you going to do the first day of your new assignment?"

I will start my first day by reporting to my command early, checking my safety equipment, reading the journal of the previous day's activities, making a proper relief, and determining if there are any planned activities for the shift. I will then meet with my command and determine if there are any SOPs I should be concerned with. I will then determine who is the cook for the shift, and then prioritize the day's activities.

What did you think of this response? If you were impressed with the fact this candidate has a practical grasp of the first day in a new assignment, you were correct. Would you really tell the interview board members you would determine who the cook is for the shift? If that is what you would really do (and you would), why not? Additionally, the word *standards* was not used, which minimized the board taking the candidate for "a ride" on a useless tangent.

What kind of a training program/fire prevention program will you utilize?

As both of these questions are similar, let's approach this from the perspective of training. Although there are various types of questions classified as important, this question (or any question in the area of training) can be classified as a *crucial sleeper* for the following reasons:

- A fire department is a success or failure depending upon its training programs. No one factor has as much ultimate affect on a department's operation as effective training (or the lack of it).

- The most important nonemergency activity in the fire service is training. The fire service is usually called first during times of various types of crisis, and is expected to take timely and definitive action to mitigate the situation. This takes constant, effective, and relevant training.

- Training yields consistency.

Without question, all activities that your command is responsible for take some type of training. Therefore, do you think it is vitally important to have a basic idea of the following:

- How you will determine the level and competency of your command?

- How you will determine necessary programs designed to maintain and improve your command?

- How you will implement those programs?

- How you will determine your training program's effectiveness and results?

- How you will adjust your training as necessary?

Your goal should be involvement in your command's activities, and to make your command the best it can be by

- Challenging your subordinates.

- Improving their performance.

- Taking your command (advancing) to a higher level than they were prior to your assignment.

Look at a few considerations that will assist in achieving that goal.

Short range.

- Determine the level, frequency, and quality of training prior to your assignment. Remember, it is possible that you have taken command of a strong company/command. If that is true, you can then build on their current strengths.

- Determine their current proficiency by

  ✓ performance at incidents

  ✓ scheduled drills

  ✓ in-house drills

  ✓ morale and initiative

- Identify strengths and weaknesses. Your goal is to build on current inherent strengths, and improve perceived weaknesses.

Long range.

- Develop goals. What do you want to accomplish and where do you want to take your command in six months, a year, etc.?

- Utilize your department's programs.

- Utilize your department's training officer.

- Utilize your expertise.

- Improve weak areas and strengthen strong areas.

- Monitor your goals and progress. Adjust as necessary.

Finally, the commandments of a training officer are listed for your consideration:

1. There are only two basic differences between a training officer and a company officer—shift duty and span of control.

2. It is imperative that the right person is assigned as a training officer.

3. View the position as a promotion, not as a demotion.

4. The job elements are not as important as investing your time wisely.

5. Seriously consider your potential impact on your department.

6. Determine your department's immediate and long range needs.

7. Determine your support and budget from the top.

8. Maximize available training opportunities (i.e., other departments, printed material, simulators, conferences, your people, etc.).

9. Enlist the support of company officers.

10. Consider the background of today's firefighter. Although new firefighters can make excellent team members, they often lack experience in the building trades, have no military experience, have never driven heavy apparatus or a car with a standard transmission, etc. These considerations dictate additional training in these areas.

11. Maximize the basics before you go ballistic.

12. Train the way you fight because your life depends on it!

13. Remember, you set the standard!

14. When possible, make your training programs interactive.

15. Require accountability.

16. You cannot gauge your effectiveness unless you can measure your progress.

This same overview or outline can also be applied to the area of fire prevention and how you would organize and administer that area.

How will you involve your company in an affirmative action and upward mobility program?

Depending on a particular area/fire department, affirmative action may or may not be a valid topic. However, upward mobility should always be a valid consideration for a supervisor. Before we look at a few considerations, you should have a definition of the terms *affirmative action, upward mobility*, and *diversity*.

- **Upward mobility.** Training for promotion. This means enhancing the promotability of your subordinates.

- **Affirmative action.** Positive effort and direction by your department to increase the representation of various persons (ethnic groups, etc.) through equal employment opportunities.

- **Diversity.** Variety. This term seems to be the current buzzword to define priority admission, cultural traditions, geographical locations, percentage plans, affirmative action, etc.

If these terms apply or are used within your department, then you should be familiar with their meaning and application to your new position. You should also be totally familiar with the goals and efforts of your department in these areas. Assuming you have a complete knowledge of your department's efforts and programs, let's look at a few items that you can affect at your level:

- **Affirmative action and diversity.**

  ✓ Determine needs.

  ✓ Know your district, city, and areas that can be successfully recruited for candidates.

  ✓ You must support and become involved in this area. Sell the program to your subordinates.

  ✓ Career day in local schools.

- **Upward mobility.**

  ✓ Consider transfers to locations that will improve member's profile.

  ✓ Utilize department career programs.

  ✓ Promote local college educational programs.

  ✓ Delegate responsibility to subordinates.

What is your opinion of the current two in/two out rule?

Although this subject is currently law, there is a significant amount of controversy regarding its implementation on the fireground. Additionally, depending on the particular department and/or area, there seems to be differing methods of implementation. You must know how your department views or requires implementation of this law, and how it will be implemented in your new position/command. However, this author believes that at the risk of revealing my dinosaur beliefs, this law has both a positive and negative approach to the fire service.

The positive side is

- There must be a minimum number of personnel (four) on the fireground prior to initial fire attack operations (unless a rescue is necessary).

- There are at least two exterior personnel when initial fire attack operations are implemented that can perform an interior rescue of attack personnel if necessary.

The negative side is

- If sufficient personnel are not on the fireground, interior attack operations will be delayed until the proper number of personnel is on-scene (however, the fire will continue to burn).

- If attack operations are delayed, building collapse and/or flashover conditions are enhanced.

- If the advance of interior attack lines is delayed by the implementation of exterior personnel, attack operations will become more hazardous.

The important emphasis of this question is that you must have your own opinion, and how you will use it in your new position.

Questions dealing with personal problems and discipline

Questions that involve dealing with people (interpersonal relationships) or some form of discipline should be anticipated and welcomed. Why? Because if you are interviewing for a supervisory position, you will be evaluated in this area, guaranteed! If you have prepared and are confident, you should welcome a question or questions in this area.

If a major portion of a supervisor's job is getting the job done (serving the people who pay your salary) through people (and it is), you must be prepared to demonstrate your abilities, styles, and methods you will utilize to assure that your subordinates maintain conformance to the standards of you and your department. Additionally, appropriate disciplinary action is an integral part of maintaining an effective and consistent fire department. The primary goal behind administering disciplinary action is to modify a member's behavior and performance.

Any firefighter who has worked for more than one supervisor is aware that there are a variety of styles and methods (affected to some degree by a specific set of existing departmental standards) that are utilized to administer varying degrees of discipline designed to achieve a desired result. Unfortunately, supervisors are either successful or unsuccessful in metering the proper amount and type of discipline for the appropriate offense.

Let's consider that last sentence from a different perspective. Depending on the offense, there are various types of discipline that will achieve the desired result, or your departmental rules and regulations will not specifically cover every possible offense. Therefore, questions that fall into the discipline category will display your ability to handle specific incidents and situations. Without a doubt, this is what the board members are looking for. How will you handle a specific situation or individual? Responses are based on various combinations.

Department guidelines. If you are given a problem involving a subordinate who has violated a specific departmental rule with a resultant penalty, your answer should parallel your department's requirements. You are being examined on your knowledge of your department's requirements. For example, let's assume your department requires a letter to the chief if a member is late in reporting to duty (whether it is 10 minutes or 1 hour). If you were asked what you would do if one of your crew reported to duty 30 minutes late, your answer would be to notify your superior when the member is late, and then have the member send a letter of explanation to the chief after the member reports to duty.

Your decision. Questions that are in this area can be summarized with the following clichés:

- "This is where the rubber meets the road."

- "Now you earn your money."

- "Bite the bullet."

If one of your subordinates needs to be disciplined and has not violated a specific departmental rule demanding a resultant penalty, you must "bite the bullet" and make a decision. In simple terms, what are you going to do? This is what the interview board is evaluating you on. Candidates faced with this type of problem usually fall into the following categories:

- Some candidates feel that an interview board is looking for something tricky—a bolt of lightning from the sky. This type of candidate tends to make the question more difficult than it is.

- Some candidates feel they must tailor their answer to what they feel the interview board members want to hear. In fact, prior to an interview, some candidates tend to look at the backgrounds of the board members to determine if they are *easy going* or *by the book*. Unfortunately, if the board members have different backgrounds, you are faced with the dilemma of trying to please everyone, which is impossible. Additionally, what difference does it make who the board members are? Either you are ready and confident, or you are not.

- Some candidates will tell the board what they will do, and stand by that decision if challenged. Obviously, this is the direction to take if you are asked how you would handle a specific situation. If you are prepared for your interview, confident of your abilities, know your department's rules, regulations, standards, steps for progressive discipline, and take a practical and common sense approach based on *getting the facts* (Fig. 5–7) before acting, you should be confident and ready to display your capabilities.

As an example of this type of question, assume you are given the following problem: "Assume you are the captain of an engine company at a structure fire. You are involved in the overhaul operation when you see one of your company members bend over, pick up a pack of cigarettes, look around, and quickly put them in a pocket. What would you do?" In this type of question, you have to make a decision based on the facts in the question. A common answer follows:

I would immediately tell my offending member that stealing is not an acceptable practice and is not conducive to enhancing the public's perception and trust of the fire service. Additionally, I would meet with the member upon returning to quarters and determine the appropriate type of discipline.

Fig. 5–7 Get the Facts!

What did you think of this answer? Let's look at several facets in this question:

✓ The fact that the member looked around and quickly put the cigarettes in a pocket has nothing to do with solving this problem. In fact, these juicy tidbits of fluff have *guilty* written all over the question.

✓ Is this member really guilty? The candidate responsible for this answer thought so.

✓ Based on a perceived verdict of guilty, this candidate verbally disciplined the member at the scene of the overhaul and then added that upon returning to quarters, additional discipline would be determined.

✓ After this answer, I guarantee the next question from the board to this candidate would be, "What discipline would you determine to be appropriate upon returning to quarters?" This candidate should then be prepared for a trip down the "primrose path" until a final decision is made.

✓ Summarizing the preceding considerations, this candidate left the door open to further questioning based on a perceived verdict of guilty. The candidate did not get all of the facts!

Let's look at another answer to the same question:

I would ask the member where the pack of cigarettes came from. If the answer was, "They were mine and fell out of my pocket as I was bending over, so I was putting them back in my pocket," *this incident would be over. However, if the member said,* "They were in the overhaul pile and appeared to be ruined so I thought I would keep them," *I would tell the member that act could be construed as stealing, is capable of creating a false image of the fire service, and not to let me catch the member doing that again. Additionally, I would make a note of the incident when I returned to quarters.*

What did you think of this answer? Again, let's look at several facets in this question:

✓ First, the company officer determined the facts. The officer in the first answer did not.

✓ Next, the candidate gave the interview board two scenarios. In the first, the cigarettes belonged to the member. End of question. In the second, the candidate was guilty as charged.

✓ When it was determined the member was guilty, the member was immediately disciplined (but not in the presence of other members).

✓ Upon returning to quarters, the officer made a note of the incident (which could be used if further similar problems are encountered).

✓ Finally, this answer got the facts, determined a course of action (from two different viewpoints), handled the problem, and minimized the interview board taking the candidate for a walk down the "primrose path."

This question (or a similar type question) is often given to see what a candidate will do—handle the problem or blow it out of proportion. The first answer assumed the member was guilty and tended to unnecessarily escalate the problem. The second answer mirrored what you would probably do if you encountered this incident. This also introduces another variable into the discipline-type question. Does the type of member (good or bad) make a difference when discipline is used, or are all members treated the same? The answer is, the type of member absolutely does make a difference (unless a specific department rule levels the playing field). I would hope that your best company member who has a perfect history, and makes a small stupid mistake is treated differently than a member who has been a constant "problem child" and makes the same small stupid mistake (that has probably been made before).

Let's look at a few definitions that should be considered in formulating a response to a discipline-type question:

Definition of discipline. Form of training to assure or maintain conformance to set standards.

Discipline is training. It is designed to change behavior.

- Positive discipline is training for motivation.

- Negative discipline is restrictions.

- Punitive discipline is dismissal, demotion, or suspension.

Utilize discipline to demonstrate that you do not condone a particular behavior.

Progressive discipline. This area is dependent on what your department utilizes. However, for discussion purposes, consider some typical steps in progressive discipline, such as:

- Informal counseling (verbal warning)

- Reduction of privileges and work assignments

- Oral reprimand

- Written notice to improve

- Performance evaluation (some departments utilize an annual performance evaluation)

- Reprimand

- Official reprimand by administrative staff

- Suspension or discharge

Documentation. The key to effective progressive discipline is documentation. Each step of your disciplinary process must be documented, and if appropriate, signed by the disciplined member. Proper documentation should include the following:

- Times.

- Dates.

- Duties involved. Who assigned the duties?

- What was done wrong? Specific standards, practices, and procedures not met. Specific rules broken.

- Training and counseling given.

Application

In summary, when determining the type of discipline, evaluate the following considerations:

- Get the facts.

- Don't be sidetracked by fluff.

- Determine if your department rules/regulations apply and act accordingly.

- If your department rules/regulations are not applicable, make a decision based on what you would do, and then tell the interview board.

- If you must use progressive discipline (it's easy if you know the steps), go through the steps *in a timely manner* with proper documentation until the problem is resolved. Remember, your job is not done until the paperwork is done.

Tips

In conclusion, allow me to suggest several "tricks of the trade" that you may find useful:

- If it is necessary to counsel a member on poor performance, the books say to first start with the member's good points and then lower the hammer! Try this. If your department utilizes annual performance evaluations, make one out in pencil (mark the good and bad). Invite the member into your office, and behind closed doors, tell the member, "If I were to give you an evaluation of your past performance, this would be the evaluation." Hand the evaluation to the member. After the member reads the evaluation, I guarantee the member will want some details.

You then compliment the member on the good points, and then discuss the points that need to be improved. Be sure to have some examples of the areas that need improvement. The key is to get the member to admit to the areas that need improvement (if you have done your homework, there are few other options), and then mutually decide on a method and an appropriate timetable to correct the area/areas needing improvement. If your department does not utilize performance evaluations, you can draft one in pencil and use the preceding method.

• From an officer's perspective, discipline is most effective if it is handled in a timely manner (the sooner the better). The longer you procrastinate the inevitable, the more it will wear on you, and the more your crew will wonder when you will "step up to the plate."

• Progressive discipline puts the ball in your court. The offending member has little to stand on if you have done your homework and completed required documentation.

• Currently, drugs and alcohol are considered a treatable disease (the fire service is not immune from these two problems). Therefore, *know* what your department policy is on this issue and be familiar with the Employee Assistance Program (EAP).

• Let's assume you are a company officer, and one of your members who is visibly upset, comes into your office and wants to discuss a problem. Try this—find out what the problem is and tell the member you will be happy to meet and discuss the member's concerns in one hour, two hours, etc., or whatever is a mutually agreeable time frame. This approach can accomplish the following:

✓ You will give the member a chance to "cool down." If possible, do not discuss problem issues when someone is upset as the mind becomes more rational when the blood pressure is not 230/480!

✓ It gives you a chance to think about the complaint and suitable solutions.

✓ If necessary, you can call appropriate persons (i.e., your boss) for advice in finding an equitable solution, and reinforce your knowledge of the rights of you and the complaining member.

• Finally, let's address one of the most challenging problems in the fire service—members who use sick time to take days off! How would you handle this problem, as most supervisors sweep this pearl under the rug? If you have a member that you suspect is abusing sick time, try this:

✓ First, determine the member is abusing their sick time. Definition of sick time: when you are off sick, you are off because you are sick!

✓ Next, make a medical calendar on the member. A medical calendar is made from the last page found on most calendars (the page that has every month with all days listed on each month).

✓ Next, mark out the days per month the member has been *sick*. Once this is done, a clear visible pattern normally emerges. For example, the member is normally off on Tuesdays and Saturdays and/or the member is off two or three days per month, etc.

✓ If a pattern is observed, call the member into your office and ask for an explanation. Be sure to use the phrase, "I notice you have been off sick an inordinate amount of days. Do you have an explanation?", or "Is there a problem I can help with?" The key word is *inordinate*. Most people cannot define the word, so you are operating from a position of strength. Additionally, as you are not a medical doctor and a pattern is obvious, an explanation is not an unreasonable request.

✓ Unless the member has a good reason (highly unlikely), this process normally cures the problem, as the member is now aware of being monitored.

✓ If this does not cure the problem, you can request a note from the member's doctor for each occurrence (this is dependent on what is allowed by your department). You can also refer this problem to your supervisors for their attention. This process will normally get results.

Can you explain the concept of conflict resolution?

The phrase *conflict resolution* is a recent addition to the fire service that you should be familiar with and able to incorporate into your command when necessary. When two or more people are in a work environment, conflict may occur. Conflict can be a disagreement, the presence of tension, or the existence of another difficulty between individuals or groups. Conflict is not a static condition. It is a dynamic process that must be dealt with to ensure a satisfactory work environment. Therefore, it is imperative that supervisors must "keep their periscopes up" and be aware of the personnel environment they are responsible for.

A supervisor's responsibility is to attempt to reduce conflict. However, when conflict does occur, a supervisor needs to take an active role in addressing the conflict and bring it to resolution. The first step in conflict resolution is to identify the problem. A common mistake is the supervisor who recognizes a small problem and allows it to develop into a significant one that requires discipline or serious conflict resolution. When conflict is recognized, the supervisor should act as a mediator and get the parties together to face one another on the issues. The goal of this initial meeting is to find mutually acceptable and long-lasting solutions. Effective confrontation resolution skills require experience and a positive constructive attitude in which the individuals are open to ideas and solutions.

Conflict definition. Conflict can be a disagreement, the presence of tension, or the existence of difficulty between two or more people. Conflict can also be a belief that "if you get what you want, I can't get what I want."

Conflict resolution process.
- Recognize the conflict and address the issue. When you act on important problems or potential problems in a timely manner, you will have more credibility—and likely, a *smaller* problem.

- Bring the parties together:

  ✓ Discuss the conflict.

  ✓ Identify the problem.

  ✓ Use active listening.

  ✓ Look for a mutual interest.

✓ Brainstorm solutions together.

✓ Be flexible on solutions and firm on interest.

✓ If necessary, indicate what you want and why you want it.

Conflict resolution rules.

- Act as a neutral moderator on a fact-finding mission.

- Use "I" statements.

- Indicate a willingness to help resolve the problem.

- Do not use the phrase, "it is your fault!"

- Stay in the present and future.

- Identify general interest areas you have in common.

- Stick to the topic at hand.

- Be optimistic.

Conflict resolution listening rules.

- Don't interrupt.

- Acknowledge their viewpoints.

- Restate what you have heard.

- Offer an apology if appropriate.

- Ask clarifying questions.

- Use silence.

Do's and Dont's

The following are dos and don'ts when answering questions:

Do

- Be comfortable; sit erect. Position yourself for good eye-to-eye contact, and use your hands for appropriate gestures.

- Be confident, enthusiastic, but not overpowering.

- Be courteous, friendly, and respectful of the board members. They are in the driver's seat.

- Use the terms; *yes sir, no ma'am, chief.*

- Listen to each question and ensure you understand each question. Do not hesitate to have the question repeated or ask for additional information if necessary.

- Answer each question promptly, but not too quickly. Take a few seconds to organize your thoughts before proceeding with an answer.

- If necessary, admit errors. However, it is difficult to retract a statement.

Do not

- Tell the board what you think they want to hear.

- Guess. If you are unable to answer a question, say so, but add that you will find the answer after the interview.

- Be negative. Problems in your department should not be openly displayed. Show loyalty to your department. Maintain a positive attitude.

- Interrupt a question. Wait until each question is completed.

- Focus on your present job. Focus on the job for which you are interviewing.

- Argue or debate with a board member. You will always lose.

- Wander. Be brief yet answer each question.

- Try to pad an answer. State the facts or your opinion.

- Answer a question with the statement; *That is a tough question.*

- Conclude your answer with the statement, *Does that answer your question?* or *I hope that answers your question.*

- Unless you know you have made a mistake, do not change an answer!

Sample Questions

The following questions are some common examples that can be expected during the body portion of an interview, and are listed in no particular order:

- What is your definition of motivation?

- What is your definition of leadership?

- In your opinion, what will be your biggest challenge in a slow fire station?

- If you are promoted to _____, what do you want to accomplish first?

- What is your strongest strength?

- What are two of your weaknesses?

- What are your goals and objectives?

- How will you implement the mission statement of this department in your new assignment?

- What do you anticipate will be your largest concern as a new _____?

- Do you have any ideas how you can increase the productivity of your new command?

- How do you solve problems?

- You are the officer on an engine company, are first-in to an accident involving a truck and a school bus. What will you do?

- You are the officer on the first-in truck to a fire, reported to be located on the 40th floor of a high-rise building. What is your size-up, how will you use the IMS, and what will you do with your company?

- What is your opinion of vertical ventilation on modern buildings?

- What is your educational background? How will it assist you if you are promoted?

- If you are assigned to administrative duty (40-hour week), what is your opinion of not being able to work on platoon duty?

- What is you definition of loyalty? Morale? Initiative?

- If you are assigned to a command that obviously needs improvement, what measures will you implement to improve performance?

- What are your future promotional goals?

- In your opinion, what is the most important element of the position for which you are interviewing?

- Where does training fit into your goals and objectives for your new command? How would you accomplish these goals? How would you determine their effectiveness?

- How would you determine the level of competency of your new command?

- What is your opinion of affirmative action and diversity in the fire service?

- Do you think the modern fire service has raised or lowered standards?

- In your opinion, why does the modern fire service consistently lose more than 100 firefighters per year? Do you have any ideas how that number can be lowered?

- What are some of the key problems facing the fire service?

- What is the difference between public relations and community service?

- How will you improve the application of community service goals within your new command?

- What were the primary characteristics of the best (and worst) officer you have worked for?

- As a new _____, what will be your most significant challenge?

- You are assigned to the position of lieutenant, and you quickly notice one of your members reports to duty just before or after the start of your tour of duty. What would you do?

- What has been the most significant benefit to the fire service in recent years?

- What kind of a fire prevention program will you run?

- What are your primary goals while conducting fire prevention inspections?

- If promoted, where would you like to be assigned? Why? Would you like to make any changes? What are they?

- In what direction is this department headed? If you could, would you make any changes? What are they?

- What improvements could you make to the current department budget?

- Can you define progressive discipline? How would you use it?

- How would you classify your use of discipline—easy or firm?

- What is your definition of the IMS? How do you use it on a daily basis?

- Within the IMS, what is the difference between an incident commander and operations?

- If you respond to a small rubbish fire that can be handled by your company, would you use the IMS?

- How would you handle an order that needs to be implemented within your command that you do not agree with? What would you say to your subordinates?

- What would you do if you suddenly discovered you had unnecessarily reprimanded one of the members of your command?

- Someone has broken into the refrigerator in the kitchen and eaten most of the food and dessert for your shift. What would you do?

- Your battalion chief is attending an important meeting in headquarters, and you are first-in to a fire on the third floor of a three-story building. What would you do until you are relieved? (*Note*: This scenario can be expanded to any type of incident you could respond to. Do your homework.)

- We notice that you only have six years in this department. Why and how do you feel you are qualified for the position of _____?

- We noticed from your application that you have 25 years in this department. Why is it suddenly important for you to receive a promotion, and how long do you expect to continue working before you retire?

- It appears some money is missing from the mess fund. What would you do?

- What is the most important aspect of truck work?

- You happen to walk out to the back of the station and notice one of your members is washing his/her car. What would you do?

- How do you establish priorities?

- What is your opinion of annual performance evaluations for you and your subordinates?

- How would you motivate a senior member of your command that is in *retirement mode*?

- What is the value of the initial size-up at a working incident?

- What do you try to incorporate in your size-ups?

- One of your company members reports to work with the obvious smell of liquor. What would you do?

- What is your opinion of a physical fitness program? Would you support one? How would you implement a physical fitness program?

- What is your opinion about conducting drills in the afternoon when your company members want to watch television and relax?

- What would you do the first day at your new assignment?

- Your female firefighter notifies you that she is being harassed by one of your members. What would you do?

- You determine that one of your members is normally gone on a trade when a significant drill is planned. What would you do?

- How would you feel if someone with a lower overall score than yours was promoted ahead of you?

- How would you handle an engineer who you thought was driving too fast to emergency incidents, but too slow to EMS incidents?

- What is your opinion of unions in the fire service?

- What importance do you place on a raise in pay?

Summary Checklist

1. The body portion of an interview will use a variety of questions to analyze your thought process.

2. Be aware of your strong and weak points. Weak points should be directed toward a positive trait.

3. Leadership is an integral part of any job. Be able to define this trait and define your leadership strengths.

4. General knowledge questions give you an opportunity to display your knowledge in current events affecting the fire service.

5. Goals and objectives allow you to gauge your effectiveness by measuring your progress.

6. Emergency incident questions will test your knowledge on standard operating procedures and/or common sense.

7. Put yourself in the seat of the position for which you are interviewing.

8. If possible, answer emergency incident problems with general concepts.

9. The fire service does not sell a product; it sells an image to the public.

10. Differentiate between public and community relations.

11. Only use the word *standards* if you are able to define your standards.

12. Training questions are important due to the fact that a fire department is a success or failure depending upon its training programs.

13. Training yields consistency.

14. Be familiar with the definition of affirmative action, upward mobility, and diversity.

15. The primary goal behind administering disciplinary action is to modify a member's behavior and performance.

16. Disciplinary questions fall into the following categories:

- Your ability to handle specific incidents

- Your knowledge of department guidelines

- You must make your own decision

17. Candidates must answer disciplinary questions from the perspective of what the candidate would do in a specific situation.

18. The key to progressive discipline is documentation.

19. Conflict can surface when two or more people are in a work environment. Conflict resolution may be necessary to solve a problem and/or disagreement.

20. A supervisor who initiates a conflict resolution meeting should focus on acting as a neutral moderator.

21. Successful conflict resolution is based on fact-finding, active listening, and finding mutually acceptable goals that are long-lasting solutions.

6. Closing the Interview

Introduction

Once time constraints have expired and/or the interview board has asked the appropriate questions and has a general idea of your interview performance, it is time for the shortest and easiest portion of an interview—the conclusion. This is normally accomplished by the interview board chairperson concluding with a final question as, "Do you have anything else you want to say or add?" At this point, most candidates (99%) will quickly thank the interview board members for their time while thinking, *If you are done, so am I. Have a nice day and I'm outta here*, and make a hasty exit.

The closing portion of an interview is necessary due to the following considerations:

- The interview board is being polite and giving you an opportunity to review any forgotten or important items not covered (Fig. 6–1).

- Interview boards must give a candidate an opportunity to have *said it all*. This minimizes the possibility of a candidate filing a protest after an interview based on the candidate's perception that there was insufficient opportunity to have said it all.

Fig. 6–1 Polite Does Not Mean Keep Talking

When a candidate realizes the interview is concluding, the first thought that comes to mind is *Is it over already?* Remember, time flies when you are having fun! However, when the interview board reaches this portion of an interview, you are suddenly faced with an important decision. Do you thank the interview board for their time and make a hasty exit, or do you give the board one last pearl designed to raise your score from 70 to 95? When an interview board reaches this portion of an interview, they are

- finished and not ready to listen to another 10 minutes of your strong points.

- familiar with your qualifications and abilities.

- fulfilling a requirement that dictates they must give a candidate a chance to have said it all.

- out of time.

- all of the above.

As there are various methods you can employ to close your interview, let's review your available options.

Missed Points

You may pick up any missed points or important items you did not have a chance to elaborate on, but this is an extremely rare occurrence. However, if you have missed any *important* points or items of consideration, utilize this opportunity to correct that deficiency. Keep it short, positive, and ensure that what you have to say is vitally important. Let me restate that last concept one more time. If you are absolutely and positively sure you missed a vital point, use the shortest time possible to state it. If you are talking more than several minutes, your grade is rapidly going downhill. Remember, the board has just indicated they are done, and the sooner you "hit the bricks," the better!

No Statement

When an interview is concluded, some candidates will quickly conclude the interview by setting an Olympic record for the 50-yard dash while leaving the interview room. Another common closing is to quickly thank the board for their time and the opportunity to share your qualifications, then leave. A common response would be similar to the following:

In closing, I would like to thank each board member for your time and the opportunity to discuss my qualifications for the position of engineer, captain, etc.

Although this response is used by many candidates, it can present a potential consideration. If one of your last answers was incorrect or not one of your better responses, this fact will still be fresh in the minds of the board members after you have left and they are determining your final grade. This has the potential to affect your grade from a negative perspective. The next option has the potential to minimize this possibility and enhance your final grade.

Closing Statement

Lastly and most importantly, let's consider an option based on the principle of *you have nothing to lose and everything to gain,* if your final response is able to adhere to an extremely specific criteria. That criteria is spelled *short* and *relevant.* It is possible to give the interview board a short, positive, upbeat statement and challenge without exhausting your welcome. This viewpoint can be used to close an interview and can accomplish the following:

- Leave the interview board with a fresh positive image of you when they are determining your final grade.

- Give the interview board a positive statement based on your demonstrated performance.

- Leave the interview board with a positive challenge relating to your future job performance.

Remember, the interview board has indicated they are finished with the interview, and they are not ready to continue for another five minutes (Fig. 6–2). Although you can leave the board with a positive challenge, it is imperative that you keep it short and sweet. This is not the time to implement the concept of *more is better.*

Fig. 6–2 Keep Closing Statements Short and Positive

Let's consider a short, positive, closing statement that meets the aforementioned criteria. Additionally, assume the following:

- Your prior performance has been rated as excellent by your supervisors.

- You are confident and enthusiastic about your qualifications and abilities.

- You look each interview board member in the eye while delivering this statement:

  Members of the board, since I have been a member of this department, my standards and level of performance have been consistently rated as excellent by my supervisors. This board can expect and depend on me to carry these same high standards and demonstrated performance into the position of lieutenant with effective results. Thank you for your time and the opportunity to discuss my qualifications.

This closing has accomplished the following:

- It will take about 12–15 seconds to complete.

- It is positive and upbeat.

- You have briefly referred to your excellent level of demonstrated performance. The phrase *demonstrated performance* means your previous performance evaluations have demonstrated your ability to perform at an excellent level. That is a justified fact. Additionally, not only is this a positive indicator of your capabilities, but if the interview board does not have access to or knowledge of your prior performance evaluations, they now know you have performed at an excellent level in your previous positions. That also is a justified fact (if you use this concept, it had better be true).

- You have given the board a promise (expect) and challenge (depend) for effective results. The words *expect* and *depend* are interesting words as used in the context of the previous statement. If you have performed at an excellent level in your previous positions, it is reasonable to conclude that you will perform at the same level in the position for which you are interviewing (expect). Additionally, and in conjunction with the word expect, you also challenged the interview board with the fact (depend) they can expect you to perform at the same high level as your previous positions. After you leave the interview room, I guarantee the interview board members will briefly consider your promise for them to expect and depend on you to perform at a high level in your new position!

At the conclusion of the interview, evaluate the following considerations:

- Immediately following your final concluding words, confidently stand and remember to firmly shake the hand of each board member as you look them in the eye (if possible, use their name and rank, if appropriate).

- If you have your department's uniform hat or cap, be sure to take it with you.

- If you moved the chair, remember to leave the chair where you found it.

- Lastly, and most importantly, if there are multiple doors, choose the right door to exit the interview room. I vividly recall an interview room having two identical doors behind the candidates.

One door went to the hallway, and the other door went to a closet. At the conclusion of a particular interview, this candidate quickly stood up, hurriedly shook our hands, turned around, confidently walked up to the closet door, opened the door and entered while closing the door behind him. About 30 seconds later, he opened the door, looked at the interview board members and said *everything seems to be OK in there*, and then exited through the correct door. Several days after the interviews were concluded, I happened to see the candidate (whom I knew). I asked him what his thoughts were when he entered the closet. This was his response:

Chief, when the door closed and it was totally dark I knew I was in trouble. I briefly considered the following options:

✓ *Stay in the closet until the board members had left at the end of the day and then leave, hoping you didn't notice that I went into the closet. But I knew you did.*

✓ *Punch a hole in the drywall and step into the hallway and then leave. But I knew this option would make too much noise.*

✓ *Just step back into the interview room and then exit the right door.*

Interestingly, although this incident was extremely humorous, it did not affect his final grade, and may have helped slightly. However, I do not recommend this approach.

Summary Checklist

1. Generally, interview boards will conclude an interview by giving the candidate time to add any points that were missed or neglected during the interview.

2. When interview boards are concluding an interview, they are finished with the candidate and ready to move on to the next candidate in a timely manner.

3. By closing with a short, positive, challenging statement, a candidate can leave a fresh positive impression in the minds of the interview board members when they are determining the candidate's final grade.

7. Preparing for a Fire Department Entrance Interview

Introduction

Although this brief chapter is not an integral part of the fire department promotional interview process, it is common for fire department members to have relatives and/or acquaintances, who are interested in becoming firefighters. This condition can result in fire department members becoming *experts* on how to become a firefighter and often results in giving advice on how to prepare and take a firefighter entrance examination. With these thoughts in mind, let's briefly consider some concepts that can be applied to assisting a potential firefighter candidate.

Normally, a firefighter candidate has to successfully compete in a portion or all of the following exams:

- Written

- Interview

- Medical

- Psychological

- Physical agility

Obviously, it is difficult to prepare for a medical and psychological examination, as you are what you are. Depending on the type of examination process, it may be possible to enhance a candidate's performance in the areas of physical agility and the written examination. However, it is relatively easy (and highly recommended) for a candidate to prepare for the interview portion of an examination process, as it is normally a major portion of the final grade that *can* be enhanced with a heavy application of applied elbow grease.

Preparation

To prepare a firefighter candidate for an entrance interview, let's quickly overview and apply the basic concepts that were discussed in the preceding six chapters:

Chapter 1

Unless a candidate is a naturally gifted speaker and also able to *leap tall buildings in a single bound*, a noteworthy amount of proper practice will be necessary.

This can be accomplished by taking practice interviews with family members, friends who are currently firefighters, etc. An essential ingredient of this recommendation is videotaping the practice interviews. Although this process can begin on the side of *ugly*, it can also develop confidence, enthusiasm, and dramatically improve a weak presentation style. Consider the following three entrance interview rules of engagement:

1. First impressions are lasting impressions.

2. Are you serious, or do you enjoy taking interviews?

3. This can be fun, because your future depends on (only) you!

Chapter 2

Entrance interviews are also comprised of an opening, body, and closing. The opening is used to start the interview with a common opening question (i.e., What can you tell us about yourself?), the body is used to evaluate a candidate in predetermined areas (i.e., decision-making ability, interpersonal relationships, etc.), and the closing is used to conclude the interview giving the candidate an opportunity *to have said it all.*

Chapter 3

Interestingly, the focal point of this chapter consisting of the four *initial* impressions a promotional candidate has the opportunity to make on an interview board are identical to the four initial impressions a firefighter candidate will also make on the interview board. Let's briefly expand on the area of personal grooming and clothing. If you were an interview board member, would it make a difference to you if a candidate were dressed in a T-shirt, jeans, and tennis shoes, or a sharp-looking suit (yes, a suit!)? The obvious difference is what the following visible messages say to an interview board:

- T-shirt, jeans, and tennis shoes give the impression the candidate is looking for "easy street."

- A suit gives the impression the candidate is professional and a serious contender.

Additionally, minimize the potential of alienating a board member with questionable grooming appearances. Stay neutral with grooming, and don't look like a Christmas tree with current body ornaments and/or resemble a poster person for a tattoo shop. Enough said?

Chapter 4

Now we get to the good stuff—preparing for the opening question, which can be similar to promotional opening questions. Some common examples are

- What can you tell us about yourself?

- What is your background?

- Where have you worked?

- Why do you want to be a firefighter?

- What are your qualifications for firefighter?

- What is the position of firefighter?

- Why do you want to work for this fire department?

To answer these types of questions, firefighter candidates can be placed into two categories:

- **Prior fire service experience.** This can consist of experience as an explorer, volunteer firefighter, paramedic, or a paid firefighter with a prior and/or current fire department. This type of candidate can discuss prior experiences specifically relating to the position of firefighter.

- **No prior fire service experiences.** Obviously, this candidate has no prior experience in the fire service and must relate past experiences that are not fire service related to future employment in the fire service. Obviously, this can be more challenging as compared to a candidate who has prior fire service experience. Most firefighter candidates are in this category.

Similar to chapter 4, consider (1) defining the position of firefighter, (2) a candidate's qualifications for firefighter, and (3) relating qualifications to the position of firefighter.

First, let's define the position of firefighter by separating the position into its basic elements (horizontal dimension). Although the position is comprised of numerous elements, the following eight attributes are worthy contenders:

1. **Dependability.** Used to demonstrate a candidate is dependable and will complete assigned responsibilities in a timely and effective manner.

2. **Adaptability to routine.** This is a subtle yet important area as the fire service demands unique variances from the *norm*:

   ✓ Other than the military, it is the only job that requires employees to eat, sleep, work, and die together.

   ✓ Workdays can be 24 hours in duration (or more).

   ✓ Firefighters will be away from their families on some holidays.

Therefore, a candidate (and family) must be willing to accept these parameters. Candidates with prior fire service experience can relate their involvement, understanding, and acceptance of these requirements. Candidates without prior fire service experience must focus on their ability to understand and accept these requirements.

3. **Character.** Examples of a character founded on respect, values, honesty, and integrity.

4. **Leadership.** Examples of past leadership opportunities (i.e., assistant manager of fast food restaurant, shift supervisor for a hardware store, etc.). Any examples of past leadership

responsibilities can be applied to the fact the candidate is self motivated, respects authority, can take/give directions, needs minimal supervision, and has been promoted to a position of responsibility.

5. **Attitude.** Examples of a positive attitude that is optimistic about the future and long-term employment in the fire service (specifically, the fire department with which the candidate is interviewing).

6. **Experience.** Examples are any fire service employment, paramedic, general prior/current work employment, and the military.

7. **Education.** Examples are high school, junior college, college, trade techs, etc.

8. **Social involvement.** Examples of involvement in community functions and organizations. Some examples are Boy Scouts, church, Little League, etc.

Because all of these attributes do not apply to every candidate, a candidate should choose four or five attributes that do apply. Once the applicable attributes have been selected, list appropriate qualifications under each attribute (vertical dimension). The candidate now has an outline of the attributes for the position of firefighter and applicable qualifications for each attribute. This can be used as an outline for a response to any of the preceding typical opening questions. For example, let's consider two responses that use attributes from the preceding eight examples and apply them to the opening question, "What can you tell us about yourself?"

First response. To answer this question, let's consider a candidate with no prior fire service experience. Additionally, let's suppose this candidate has a background resulting in the following resume:

- **Social involvement**

    ✓ member of Little League team

    ✓ Boy Scout Council

- **Character**

    ✓ integrity

    ✓ optimistic

- **Dependability**

    ✓ respects authority

    ✓ punctual

- **Experience**

    ✓ assistant manager fast food restaurant

    ✓ rental company

- **Attitude**

    ✓ positive

    ✓ willingly accepts challenges

Now, let's apply this resume to the opening question, "What can you tell us about yourself?" (Fig. 7–1)

Entrance Interview Sample #1

Members of the board, I am 23 years old and am currently attending Jones College majoring in fire science while simultaneously working for a large rental company. However, I desire to make my career in the fire service, and specifically, this fire department. Allow me to take several minutes and describe the position of firefighter and my qualifications for firefighter.

The position of firefighter is comprised of five major attributes. They are character, dependability, experience, attitude, and social involvement.

In the area of character, I believe that my focus on integrity will allow me to be a positive and consistent example to my coworkers and the citizens of this community. My strong optimism for the future is strengthened by the prospect of making a career with this fire department and being able to become an integral part of this team.

In the area of dependability, the combination of my prior work experience and college responsibilities have demonstrated my ability to respect and follow the directives of my supervisors. Additionally, I understand the importance of being punctual, timely, and accurate with assigned tasks. As a firefighter, I will use these traits to allow my supervisors to depend upon me as a firefighter who can be relied on to complete assigned tasks in an efficient and timely manner.

In the area of experience, my employment with a large rental company has given me the opportunity to maintain and operate various kinds of mechanical equipment. As a firefighter, this experience will enable me to better understand, maintain, and operate various types of mechanical equipment, both from an emergency and nonemergency perspective.

Fig. 7–1 Entrance Interview Sample #1

The area of attitude is an important qualification for a firefighter. A person's attitude is often the foundation for the outlook and approach to life and a career. As a firefighter, this department can expect a positive and goal oriented attitude that is focused on excelling in the numerous challenges in the fire service.

Finally, in the area of social involvement, it is essential for a firefighter to be committed and involved in the fire service. However, it is also important to be involved within the community you are sworn to protect. My involvement in a Little League team and the Boy Scouts of America has strengthened my vision, that as a firefighter I will represent this fire department to the citizens of this community in a friendly and professional manner.

In summary, this board can expect and depend upon me to utilize these same qualifications, attributes, and personal traits as a firefighter with this fire department.

Fig. 7-1 *Continued*

With this response, notice the following:

- The candidate answered the question in the first sentence of the response then quickly overviewed the primary goal of the interview. (The candidate desires to change careers and become a firefighter.) Then the candidate offered to tell the board what the position of firefighter encompassed and related appropriate qualifications. From a board member's perspective, what a deal!

- The candidate then described each attribute of the position of firefighter and related applicable qualifications. Notice that due to the fact this candidate had no prior fire service experience, some of the qualifications are *implied* when related to the position of firefighter. Remember, depending on the type of qualifications, this type of candidate may not be able to *directly* relate prior fire service experience to the attributes of a firefighter.

Second response. Now, consider the same question, "What can you tell us about yourself?" and consider an answer from a candidate *with* prior fire service experience. Additionally, let's suppose this candidate has a background resulting in a similar resumé to the preceding resumé but incorporates the attributes of fire service experience and adaptability to routine as follows:

- **Social involvement**

  ✓ member of Little League team

  ✓ Boy Scout Council

- **Character**

  ✓ integrity

  ✓ optimistic

- **Dependability**

  ✓ respects authority

  ✓ punctual

- **Experience**

  ✓ volunteer firefighter

  ✓ construction company

- **Adaptability to routine**

  ✓ fire service experience

  ✓ Air National Guard

Now apply the same opening question to this resumé (Fig. 7–2).

Entrance Interview Sample #2

Members of the board, I am 23 years old and am currently attending Jones College majoring in fire science and working for a construction company, serving in the Air National Guard, and participating as a volunteer firefighter. However, I desire to make my career in the fire service, and specifically, this fire department. Allow me to take several minutes and describe the position of firefighter and my qualifications for firefighter.

The position of firefighter is comprised of five major attributes. They are character, dependability, experience, adaptability to routine, and social involvement.

In the area of character, I believe that my focus on integrity will allow me to be a positive and consistent example to my coworkers and the citizens of this community. My strong optimism for the future is strengthened by the prospect of making a career with this fire department and being able to become an integral part of this team.

In the area of dependability, the combination of my prior work experience and military responsibilities has demonstrated my ability to respect and follow the directives of my supervisors. Additionally, I understand the importance of being punctual, timely, and accurate with assigned tasks. As a firefighter, I will use these traits to allow my supervisors to depend upon me as a firefighter who can be relied on to complete assigned tasks in an efficient and timely manner.

In the area of experience, my employment with a construction company has given me the opportunity to work in building construction and to better understand how buildings are built and how they can fail. As a firefighter, I will be able to use this knowledge in evaluating structural fireground incidents. As a volunteer firefighter, I have had the opportunity to respond to emergency incidents, accompany firefighters on fire

Fig. 7–2 Entrance Interview Sample #2

prevention inspections, and assist with station and apparatus maintenance. As a firefighter with this department, my familiarity with these various areas of firefighter responsibilities will enhance my ability to quickly become a member of this department's team and a dependable effective firefighter.

The area of adaptability to routine, demands that candidates are willing to accept the requirements of a fire service career. As each of you are aware, the fire service requires that firefighters eat, sleep, work, and possibly die together on a 24-hour basis. The opportunity to serve as a volunteer firefighter has enabled me to successfully work within these parameters. As a result, I can accept these requirements and inherent risks and look forward to making this my life career.

Finally, in the area of social involvement, it is vitally important for a firefighter to be committed and involved in the fire service. However, it is also important to be involved within the community you are sworn to protect. My involvement in a Little League team and the Boy Scouts will represent this fire department to the citizens of this community in a friendly and professional manner.

In summary, this board can expect and depend upon me to utilize these same qualifications, attributes, and personal traits as a firefighter with this fire department.

Fig. 7–2 *Continued*

The major portion of this response is the same as the preceding response with the following exceptions:

- In the area of experience, the candidate can relate prior fire service experience. A candidate with this background can discuss being familiar with the fire service, responding to emergency incidents, importance of station and apparatus maintenance, training, representing the department to the community, etc.

- In the area of adaptability to routine, this candidate can directly relate to the fact of familiarity and acceptance of the unique requirements of the fire service. This is a strong positive for a candidate, and is not easily duplicated by candidates with no prior fire service experience (candidates with military service can relate the same type of knowledge and acceptance).

Chapter 5

Once the opening question has been asked and answered, the interview board will then ask predetermined questions that assist the board members in evaluating the candidate. Generally, these questions are simplistic and general in nature, as most candidates have no prior experience in the fire service. These questions must then be answered from the applicant's background and ability to solve common sense problems.

Additionally, it is *strongly* recommended that any fire service candidate familiarize himself/herself with a general knowledge of the fire department for which they are interviewing. Examples are the number of fire stations, size of the department, pay, benefits, mission statement, work schedule, type of apparatus, etc. What would your opinion of a candidate be if the candidate knew nothing about the department being interviewed for as opposed to a candidate who was familiar with the preceding examples?

Sample Questions

The following questions are common examples that can be expected during the body portion of an interview and are listed in no particular order:

- What will your immediate family think of your becoming a firefighter?

- What are your strengths and weaknesses?

- How do you feel about living and working with coworkers for 24 hours or more?

- Why have you chosen this particular fire department for employment?

- How do you feel about living with male or female coworkers?

- Do you have any noteworthy talents?

- How do you feel about preparing meals for your company members?

- You notice one of your fellow company members periodically sneaking outside the fire station, and you accidentally determine that drugs are involved. What would you do?

- Do you have any goals?

- Do you have any other applications on file with other fire departments? What would you do if you were called to this department and another fire department at the same time?

- Your company officer orders you into a building that is on fire and you determine (or think) it is an unsafe operation. What would you do?

- If you had to choose between rescuing a firefighter or a civilian inside a burning building, who would you rescue and why?

- While on the fireground at an incident, you notice company members are using firefighting equipment with different methods than you were taught at the drill tower. What would you do?

- You determine one of your fellow firefighters is working on days off. What would you do?

- What do you know about this fire department, and why do you want to work here?

- What do you intend to do with your spare time around the fire station?

- How do you feel about minority hiring quotients and goals practiced by this department?

- While at a fireground incident, you happen to see one of your coworkers pick up some jewelry, look around, and quickly put it in a pocket. What would you do?

- Do you, or have you ever had any relatives in the fire service?

- Which of your prior jobs have you liked the most, and the least. Why?

- Do you have any work experience that will be useful in the fire service?

- What is the difference between an engine, truck, and rescue?

- Do you have any future plans with this fire department—education, promotions?

- Your company officer directs you to take a hoseline to the back of a structure. While doing so, another company officer directs you to take the hoseline to the front of the same structure. What would you do?

- The fire service is one of the top three most hazardous jobs in this country. In fact, more than 100 firefighters are killed in the line of duty each year. How does that affect you and/or your family?

- Why have you chosen this fire department as a potential place of employment? Do you plan spending your entire work career with this fire department?

- Are you familiar with the benefits offered by this fire department? What are the negatives if you work for this department?

- How do you work with people who have different interests than you?

- What other duties are firefighters responsible for besides extinguishing fires?

- There are times when firefighters must deal with unpleasant incidents, as auto wrecks, tragic medical incidents, and death to civilians or other firefighters. What is your opinion of this fact and how would this affect you?

- Tell us about your last job. Why did you leave (or are thinking about leaving)?

- Have you ever been fired from previous employment? Why?

- What don't you like about the fire service?

- If you have any spare time in your new assignment, how will you use it?

- What is the most important attribute of a firefighter?

- How many previous exams have you taken for employment in the fire service? Why were you not hired?

- What do you think of unions in the fire service?

- Why are you the best-qualified candidate (or, why should we hire you)? What can you offer this fire department?

- How can you improve this fire department?

- What are your main hobbies and/or interests?

- After dinner, you notice the smell of liquor on the breath of your company officer. What would you do?

- Have you ever worked for an organization that is militaristic?

- You are suddenly notified that your relief is sick and it will be necessary for you to work additional hours. What would you do if you had an important commitment that same morning?

- Are you involved in any community functions?

- You are a new probationary firefighter, and you notice your probationary relief on the other shift is not accomplishing assigned responsibilities. What would you do?

- How do you intend to handle periods of extreme stress?

Chapter 6

When the interview is concluded, the board will ask the candidate, "Is there anything else you would like to say?" Similar to chapter 6, there are three options. The preferred option is to give the interview board a *quick* final closing leaving the board members with a positive image of the candidate while the final grade is being determined. An example of a brief closing is:

During my prior work history, I have developed a positive respect for my employers that has resulted in a strong commitment to excel in my work ethics and standards. This interview board can expect and depend upon me to carry these same standards and ethics into the position of firefighter with this department. Thank you for your time and the opportunity to discuss my qualifications.

Summary Checklist

1. In most areas of this country, competition for the position of firefighter in the fire service can be best described as highly competitive, or candidates with the best balance of preparation and background are serious contenders. Only the correct application of applied elbow grease can transform a firefighter candidate from a participant into a serious contender.

2. Try to keep opening responses (to the initial opening type question) under four minutes.

3. If a closing is used, do not exceed about 15 seconds.

4. Immediately following the final concluding words, confidently stand and remember to firmly shake the hand of each board member and look each one in the eye, using each one's name and rank if possible and appropriate.

Index

A

B

C

D

E

F

G

H

I

J–K

L

M

N

O

P

Q

R

S

T